This book is to be returned on or before
the last date stamped below.

RICHARD
MABEY

PLANTS WITH
A PURPOSE.

RICHARD MABEY
PLANTS WITH A
PURPOSE.

ST. JAMES'S HOSPITAL

Plants with a Purpose

By the same author

FOOD FOR FREE
THE UNOFFICIAL COUNTRYSIDE

Richard Mabey

Plants with a Purpose

A guide to the everyday uses of wild plants

*With 8 colour plates and
79 line drawings by*

MARJORIE BLAMEY

COLLINS
St James's Place, London

William Collins Sons & Co Ltd
London · Glasgow · Sydney · Auckland
Toronto · Johannesburg

First published 1977
© Richard Mabey 1977
ISBN 0 00 219117-2
Printed by Jolly & Barber Ltd, Rugby
Bound by W & J MacKay Ltd, Chatham

Contents

Contents

For Fran

Acknowledgements

My thanks are due to *The Countryman* magazine, and to Kathleen Hersom, for permission to reprint parts of her article 'Games with Flora' I am grateful for the sharp eyes and skilled hands of many friends and helpers, who supplied me with a steady flow of out-of-the-way information, and experimented with the ideas and recipes in the book – Nigel Ashby, Marc Constantine, Tony Evans, David and Gillian Mabey, Mary Miller, Peter Newmark and Elizabeth Roy. Particular thanks to David McClintock, who corrected my erratic plant nomenclature, to Sarah Hobson for her herbal oils, to Robin McIntosh, who helped with much of the research, typed the manuscript and prepared the index, and to Francesca Greenoak, for her exquisite pot pourris and lavender bottles, her wise comments on early drafts, and her sensitivity towards wild flowers, which was and is a source of constant inspiration.

List of Plates

References to the full range of uses of each plant illustrated can be found in the index.

Introduction

In the early 1970s, two Australian scientists succeeded in unravelling the source of that fleeting and evocative smell that follows a fall of rain on the earth. Before starting on their search, they had christened the unknown aromatic 'petrichor' – the essence of stone. And essences in stone is what it proved to be, the fragrant oils and resins that are released by green plants in sunshine, washed down by rain, absorbed by clays and porous stones, and then released by the next rainfall. They calculated that the combined weight of the volatile substances released from the world's seeds, leaves and barks could amount to over 400 *million* tons each year. And with this extraordinary figure in mind, they were able to suggest that these oils might, over the immense lengths of geological time, have laid down many of the petrochemical deposits of the hot regions of the earth.

The smell of rain, carried impalpably in the summer breezes and absorbed as solidly accountable chemicals in the rocks – we're not used to such harmonious coincidences between commerce and romance. (Though Indian merchants were wise to this one long ago, and in the area round Lucknow, market a perfume known as *matti ka attar* – 'earth scent' – made by distilling sunbaked clays with sandalwood oil.) Yet for thousands of years before the advent of mass-production, articles made from natural products tended to be as pleasing to the senses as they were efficient at their jobs. Sometimes this was a result of deliberate 'artfulness' by the craftsmen. But more often it was a natural consequence of the immense range of smells and shapes and textures to be found in the plant world, which plant craftsmen wisely took as the guidelines for their designs. It would be difficult, for instance, given the variety of shade and shape of willow twigs, for a wicker basket not to be aesthetically attractive; or for a posy of moth-repellent meadow flowers for the laundry basket not also to be pleasing to the eye and nose.

It was not so long ago, of course, that wild plants were the chief raw materials in household economy. Buds, roots, flowers, sap, were all pressed ingeniously into service. Hazel branches were used as the foundation for walls, and reed thatch for roofs. Woods were carved into spoons and plates, and the shavings used to stuff mattresses and kindle fires. Twigs were bunched into brooms and woven into baskets. An immense variety of herbs and berries were used to treat everyday ailments – because doctors, like professional builders and furniture makers, were remote and expensive; and, in any case, an understanding of how to use plants – 'wortcunning' as it was once delightfully known – was regarded as a natural and essential domestic skill.

This book is an attempt to revive some of those skills, and to describe the ways in which the commoner wild plants of Europe and North America have, and can, be put to fruitful and enjoyable household use. In no way does it pretend to be a comprehensive guide to all our useful wild plants. It would be possible to make such a catalogue, to tabulate the uses of holly and honeysuckle and honewort (Gerard reckoned even this rare plant had at least one, in reducing a type of facial swelling known as a 'hone'). But such a list would not only be almost endless, but also endlessly repetitive, for so many of our plants have been used in similar ways. So I have concentrated instead on a representative selection of plants that illustrate *styles of approach* to plant use, and particularly on those plants – like the broom and the teasel – that are so tailored for certain functions that they have lent their name to the products and processes associated with them. 'Seeing' such potentialities in a plant – a walking stick in the angle of a branch, an ink in the black deliquescence of a toadstool – is what I would hope to encourage by this book; and why, even though it relies on a good deal of evidence from the past, it is really about possibilities for the future. Wild plants, almost by definition, are those whose economic potential has not yet been explored sufficiently for them to have been taken into cultivation. (Or, whose exploration has been unnaturally cut short because of commercial pressures.) There are something like 300,000 species of plant in the world, and it is reckoned that no more than four percent of these have been thoroughly investigated for useful products. The quantities of undiscovered drugs and materials – and infinitely renewable, non-polluting materials at that– must be incalculable. Yet we are still destroying the virgin forests where so many of these species have their last remaining habitats; and with them, their human inhabitants, who in their long and intimate partnership with plants have unlocked so many of their secrets.

There are not many plants still undiscovered in the temperate zone, but we too have lost that sense of intimacy. Making simple plant-based articles is one way of beginning to reestablish that closeness, and to sharpen awareness of our dependence on the plant world. Not that all the benefits are so abstract. In an age of interminable winter power cuts, a rushlight (p. 134) can be an exceedingly practical device. If you wish to shine up your furniture, a walnut will do the job more cheaply and with a more delicate fragrance than most proprietary polishes. And the discovery of the insecticidal and antibiotic properties of an increasing range of plants (p. 86) is a crucial area of investigation in which any gardener or naturalist can play his part. Even by making a simple walking-stick for yourself, and following through the observations on shape and durability that our ancestors must have made, you may begin to develop that eye for the job that has been at the root of all imaginative plant use.

In earlier times that kind of perception was inseparable from magic. If we're apt to poke fun at our ancestors' beliefs about the workings of nature, it is as well to remember the kind of world they lived in. Eleanour Sinclair Rohde has given a graphic description of what it must have seemed like to a Saxon peasant:

. . . an epoch when man fought with Nature, wresting from her the land, and when the unseen powers of evil resented this conquest of their domains. To the early Saxons those unseen powers were an everyday reality. A supernatural terror brooded over the trackless heaths, the dark mere pools were inhabited by the water elves. In the wreathing mists and driving storms of snow and hail they saw the uncouth 'moor gangers,' the 'muckle mark steppers who hold the moors,' or the stalking fiends of the lonely places, creatures whose baleful eyes shone like flames through the mist. To this day some of our place names in the more remote parts of these islands recall the memory of those evil terrors.

The Old English Herbals, 1922

No wonder that some system was developed to make sense of these fearful and dimly comprehended forces, and of plants that could, apparently capriciously, feed you, kill you, or send you mad. Though it may have lacked the rigour of our modern experimental methods, it had its own rules and techniques. It was based on the principles of sympathy and antipathy, and on a search for pattern and association. Its predictions about the power and properties of things were based on the way they behaved in other situations. It was applied most extensively in the field of medicine, for disease was rampant and terrible, and perhaps the most bewilderingly invisible of all processes. So a plant like parsley-piert, whose roots grew through gravel and stone, was believed to be equally effective at breaking through kidney stones, if taken into the body. In an age when man felt himself to be part of nature, not its master, it was surely not such an unreasonable thing to believe that the way plants grew in the earth would be reflected in their actions upon the body.

This was our ancestors' science. We deride it as magic – though if it hadn't claimed a basis in 'truth' we might applaud it as poetry. Later many of its practices were suppressed by the Christian church, who were then as opposed to any whiff of egalitarianism in the relationships between man and brute nature as they were to the pagan rituals by which the two reached some sort of working compromise. The remnants of the old magic were separated from their roots in working communities (where experience and common sense usually whittled down the more outlandish excesses) and appropriated by priestly and professional elites. Finally they were taken over by charlatans like Coles and Culpeper, and gave birth to the concepts like the infamous Doctrine of Signatures. This decreed that plants revealed their usefulness in medicine by their shape or colour. So walnut kernels were given for disorders of the brain, and yellow flowers for jaundice.

In spite of the extravagant nonsense of many of these beliefs, they did at least preserve something of the whole-hearted human response to plants that had characterised that earlier era of ecological magic. They upheld a respect for plants, a joyousness in their forms and benefits, and the principle that imagination, as well as necessity, is the mother of invention. It was, after all, an open-minded investigation of the potions of a Shropshire herb-witch in the late 18th century that led William Withering to uncover the effects of the foxglove on the heart. The powdered leaves, as *Digitalis*, are still the most valuable drug in the treatment of heart failure.

If imagination is a vital factor in exploring plant use, so is a sense of ingenuity and thrift. Nature is of necessity entirely efficient in its use and recycling of raw materials. Green plants are the only agents we have on the earth that can directly transmute the light of the sun into solid matter. Left to themselves, or husbanded respectfully, they will do this every year without cease. Their remains are non-polluting, biodegradable, and go back to the earth to provide the nourishment for their successors, obeying what John Stewart Collis called 'the law of return'. Nothing in nature is wasted. With our limited mineral resources fast being squandered, and the earth poisoned by the remains of synthetic chemicals that are unknown in nature and do not follow that 'law of return', the crucial importance of plants can no longer be ignored. They are not just vital sources of energy and raw materials, but models of harmonious self-sufficiency in their own right.

This sense of thrift is followed by the best plant-craftsmen. Some groups of the Paiute tribe in Nevada have built a whole culture around the cattail, or reedmace. They pick the submerged shoots as their first green vegetable of the spring. They weave the broad leaves into boats for catching ducks and collecting eggs from the reedbeds. The pollen from the flowerspikes is turned into bread, the dough, typically, being wrapped for baking in cattail leaves.

When the flowerspikes have seeded, the grains are roasted and made into a gruel, and the down that surrounds them used as a stuffing for cradles and mattresses.

Six thousand miles away, in the beechwoods of southern England, the bodgers followed the same frugal principles. These carvers of chairlegs, who lived and worked in the woods for a whole season, built their temporary huts from their first cuttings of wood, insulated them with the shavings that accumulated beneath their benches, and, when they had cleared their patch of trees, cut up the benches themselves to provide a final batch of legs.

The old English hedge is perhaps the supreme example of the prudent economies practised by an earlier generation of workers with plants. There was a remarkable book published in the middle of the 18th century, entitled the *Compleat Body of Husbandry*, which, unlike most other farming books of this period, was not written by some pontificating landowner, but by a team of journalists reporting on the practices being followed in working farms throughout the country. Its remarks on hedges are highly relevant today. Although hedges were principally set as boundary markers and stockproof barriers, such considerable strips of growing wood were too useful to have just a single function. So they were cropped, as if they were linear woods, and used in a variety of domestic crafts:

When the hedge is made of whitethorn, the careful husbandman will not think it finished when he has set that alone; he will plant in it, at proper distances, timber, or fruit trees . . . and they thrive as well as in orchards; elsewhere it may be proper to plant crabs, and pear-stocks, for the use of the orchard, in grafting apples and pears. Even the whitethorn is not without its use beside that in the fence; for its root, when of a certain age, is knotted and veined in a most beautiful manner, and serves the cabinet-makers for many of their elegant works . . . Mr Ellis, who is a person of veracity, affirms from his own knowledge, that a farmer in Hertfordshire, who occupied only sixty acres of land in enclosed fields, made in one season a thousand faggots from his hedgewood alone.

Nowadays our hedges are not so much harvested as emasculated, and whitethorn, crab-apple and young oak alike are reduced to shreds by the flail-mowers. No wonder a Wintu Indian woman from California was moved to say:

The White people never cared for land or deer or bear. When we Indians kill meat, we eat it all up. When we dig roots we make little holes. When we build houses, we make little holes. When we burn grass for grasshoppers, we don't ruin things. We shake down acorns and pinenuts. We don't chop down the trees. We use only dead wood. But the White people plough up the ground, pull up the trees, kill everything. The tree says 'Don't. I am sore. Don't hurt me'. But they chop it down and cut it up. The spirit of the land hates them.

Quoted in *The Challenge of the Primitives*, Robin Clarke & Geoffrey Hindley, Jonathan Cape, 1975

Raw Materials and Basic Techniques

We have probably become too civilised ever to adopt the North American Indians' religious sense of unity with nature. But we can follow their code of behaviour towards plants as a set of sensible and practical conservation principles: 'When we kill . . . we eat it all up'. Regard apparently wasted plant stuff as the basic source of your raw materials. See your gathering as a form of gleaning, and cut fresh plants only when it is absolutely necessary. Look for wood that has been broken off in storms, and for the abandoned clippings of hedges. It is alright to cut small quantities of twigs, rushes, leaves and berries (using a knife to avoid damaging the parent plant), but if you wish to cut larger branches, ask the permission of the tree's owner. Only pick flowering plants where they are growing in profusion, and remember that in Britain, at least, it is an offence to uproot any wild plant that is not on your own land – though your garden alone should supply plenty of wild 'weeds' that can be used for the products described in this book. And be thrifty about the by-products of your work. Bark strippings can be used for dyes, and the fluffy seed heads of reedmace as stuffings.

There are five broad categories into which almost all the products of wild plants can be placed: their chemicals, for medicines; their colours, for dyes; their scents; their woody parts, for carving; and their fibrous and twiggy parts for weaving. Examples of the use of these products crop up throughout the book in entries on specific plants and articles. But, as subjects in themselves, they are vast topics, too big to be included here, and I would refer readers to the books on these topics in the bibliography. I have confined myself here to some general introductory notes on these five basic classes of use.

Medicinal herbs

Plants are immensely complicated chemical factories which can turn the relatively simple ingredients of air and water into enzymes, sugars, proteins, solid cellulose, liquid oils, scents to attract pollinating insects and poisons to kill off predators. No wonder that what are arguably the four most important drugs in the history of medicine all derive directly from plants. Quinine comes from the bark of a Peruvian tree; morphine from the seed capsules of the opium poppy; penicillin from a common kitchen mould. Even the Pill would be a commercial impossibility if some of the complex chemicals in south American yams were not available as starting materials.

But I did not feel that this book was the place for any lengthy discussion of herbal medicine (the treatment of disease with unadulterated plant substances). I have tried to write about the pleasures of using plants, about the relation between function and satisfaction, and most of the grim potions of the herbalists do not belong in this kind of celebration. Where herbal remedies *are* enjoyable (see, for instance, BALSAMS, ELDER, HOP PILLOWS, MARSH-MALLOWS) I have been more than happy to include them.

This is not in any way to underrate the value of plant-based drugs, which, as I have indicated above, are still one of the cornerstones of conventional medicine. Over fifty per cent of contemporary prescriptions are still for medicines which contain plant derivatives, and even our native flora has given us remedies as diverse as aspirin for the head (derived originally from willow bark) and colchicine for gouty feet (from the root of the meadow saffron).

The other difficulty is that the effectiveness of many of the less orthodox herbal remedies is, to say the least, open to argument. Birch brooms and teasels work for everyone. But lichens scraped from a crucifix (for nightmares), and shepherd's-purse-and-cobweb ear plugs (still in herbals in the 1970s) are more selective in their beneficiaries. Every herb has its zealous advocates, and witnesses to its role in miracle cures. Rudyard Kipling once wrote that:

> Anything green that grew out of the mould
> was an excellent herb to our fathers of old.

If he had been more cynical he might have added that every herb was also, if you considered enough points of view, an infallible panacea. In one esteemed herbal handbook written as late as 1966, wild and garden carrots are guaranteed to relieve the following disorders: bad wounds, old sores, swellings, tumours, cancer, anaemia, jaundice, diabetes, internal ulcers, worms, styes in the eyes, blindness, kidney and bladder troubles, dropsy, lymphatic disorders, painful menstruation and varicose veins.

To be fair the carrot *is* a valuable medicine. It contains large quantities of vitamin A, and at a charitable estimate a fifth of the afflictions above could be due, in some cases, to vitamin A deficiency. But would that such a formidable list of scourges could be taken care of by such a simple matter of diet!

For myself, I would prefer to leave diagnosis up to qualified doctors, encourage pharmacologists to search ever more urgently for safe (but genuinely effective) natural drugs, and to give the old herbalists the benefit of the doubt by guessing that they *did* often cure their patients – not by magic, but by a combination of luck, suggestion and the addition of badly needed green vegetables to their usually poor diet.

I am completely in sympathy with the late Euell Gibbons (whose book *Stalking the Healthful Herbs* is the most level-headed and sensual romp through folk medicine I know), when he said that if he felt the need for a herbal medicine, he preferred eating it as an appetisingly cooked vegetable to gulping it down as a pill or rancid infusion. I personally find a good sniff an equally agreeable method, but the principle is the same. Let me add some brief notes on a few plants that are both beneficial and tasty. Every one of these plants is quite harmless, and of agreed efficacy by both orthodox physicians and herbalists.

As a tribute to Euell Gibbons, I feel I must give pride of place to white horehound, one of his favourite plants, and long valued as a remedy for sore throats. *Marrubium vulgare*, a furry-leaved member of the mint family, is a rather uncommon plant of dry and waste places in Europe, and widely naturalised in the United States. The leaves have a warm, musky smell and make a rather bitter tea. In medicine they are usually converted into sweet lozenges, made by boiling an infusion with sugar until it is sufficiently concentrated to solidify on cooling. But how much more satisfying if you could find wild horehound candy ready-made. Euell Gibbons describes how one of his neighbours produced this exceedingly tasty medicament by one of the most ingenious natural processes I have ever come across:

I know a man who keeps a hive of bees in the middle of a patch of wild horehound. When the horehound starts blooming, he removes all accumulated honey from the supers where bees store surplus honey; then when the plants have finished blooming, he takes out the honey that has been made from the horehound blossoms. The taste and smell of horehound is quite noticeable in this honey. When any of his family have a cough, he gives them a tablespoon of this natural cough sirup at bedtime.

Stalking the Wild Asparagus, McKay, 1962

The addition of honey is a sure way of making many herbal tisanes palatable. But you will not need it with *tilleul*, a tea made from lime flowers, which are quite adequately furnished with their own honey, and are a recognised mild sedative; or with mint or chamomile teas, both valuable in the treatment of indigestion. Fennel seed is also an officially approved carminative, which is why they serve it after curries. Once in an Indian restaurant I was given an after-dinner chewing mixture which consisted of fennel seed, betel root and patchouli-scented sweet balls, and which made every part of my insides from teeth to stomach feel as if it had been freshly laundered.

If you need a tonic to encourage your appetite (rather than something to remedy its excesses) try a salad of dandelion leaves. These contain a bitter principle called taraxacin, which as well as stimulating gastric secretion, acts as a mild laxative. They also contain almost as much vitamin C as oranges (though rose-hips contain nearly twenty times more.) Most impressively they contain as much Vitamin A as that celebrated source, the carrot. So, if you persist with your dandelion hors d'oeuvres, maybe you will be cured of those twitching eyes and old sores and varicose veins. . . .

Dyes

Natural dyes are amongst the first useful products that come to mind when wild plants are thought of, yet I have covered very few of them in this book. It is true that the colours they produce are soft, subtle, lustrous, aromatic, and best of all, alive. They change and move not just through the course of the fabric, but through the passage of time. At the same time they are troublesome to apply, nothing like as fast as synthetic dyes (often needing expensive toxic chemicals to fix and bring out their brilliance), and demand prodigious quantities of fresh plants in their production. For any reader who believes that this apparent folk-art has no links with high industrial technology, let me give this account of the manufacture of the most venerable of all plant dyes, the woad of ancient Britain.

The leaves of this member of the cabbage family, which is native to southern Europe, were spread and dried in the sun, then ground up in a mill to a paste. This was formed into heaps exposed to the air (but protected from the rain) until the pulp started to ferment. A crust formed over the piles, and great pains were taken to ensure that this did not split. But when fermentation was complete, the paste was once more pulped and formed into cakes. Before being used by the dyer, these cakes had to be broken up, moistened and fermented yet again. A hundredweight of leaves was needed to produce ten pounds of the final dye, and the stench of the fermentation was so appalling that Queen Elizabeth I ordered woad production to be halted in any town through which she was passing. How the ancient Britons managed such sophisticated techniques in the production of their legendary war-paint is a mystery. (There is, incidentally, a fitting sequel to its ancient use in the combatative arts: it was last employed in Britain in the dyeing of policemen's uniforms.) Yet the complexity of woad processing is by no means untypical, and prohibitive weights of almost all wild plant material are needed for dyeing fabrics or hanks of wool of any appreciable size. At one time, in the Outer Hebrides, so much lady's bedstraw was being pulled up for the red dyes in its roots that grazing lands were being seriously eroded, and the practice had to be forbidden.

In addition, most plant dyes need the addition of metallic salts like alum or tin chloride to bind the colours fast to the fibres, and prevent them from fading. There seems little point to me in replacing synthetic dyes with plant products if you need to bolster them up with the highly energy-consumptive and often poisonous products of the chemical industry.

Nevertheless natural dyes are too pleasant to ignore, and if you are prepared to enjoy rather than resent the delicate fading that results when you don't use metallic fixatives (mordants), and to confine yourself to items like scarves and socks, you will only need very small quantities of plant material. Almost any plant will colour wool or cloth, and you should experiment to find which ones you enjoy the most. Reed flowers, for instance, give a light green, pine cones a reddish yellow and bracken leaves a brown. Privet carries two potential colouring substances, a yellow in the leaves and a bluish green in the berries. But the alder tree boasts four: a tawny red in the bark, a pinkish fawn in the fresh greenwood, a green in the catkins, and a yellow in the young shoots. The dark colour from gipsywort was even used as a primitive sun-tan lotion by early gypsies wanting to pose as 'swarthy Africans'. I have given a list of eight commonly available dye plants which will give fairly fast colours without chemical fixatives in the captions to Plate 6, together with brief instructions on their use.

But as a general set of rules for using dye plants without mordants, it should be enough to keep to the following principles. Wash and clean the plants to remove any dirt or dead leaves. Simmer them in water in an enamel or steel saucepan until you have the colour you want. Berries will give up all their colour within minutes, but barks may take hours. As to the quantities to use, about 100 gms of plant material to 5 litres of water will do for most small jobs – though I prefer the rule-of-thumb custom of an old Kentucky dyer who said 'I pays no attention to the amount. I just throws it in until it looks right'.

Allow the dye liquor to cool, strain, and then insert the item to be dyed (which must be a natural fibre like wool or cotton, and which must previously have been well washed and rinsed). Leave it in the liquor for as long as it needs to reach the required colour, overnight if necessary. Then rinse once in hot water, once in cold, and hang up to dry.

If you wish to perform the entire dyeing sequence in one operation, pack the dye-plant into a stocking or muslin bag, put it into a large pot together with the article to be dyed, and simmer until you have the colour you want. Then treat the fabric as if it had been dyed separately.

The most specialised class of naturally fast dye plants are the lichens. But as I have explained on page 110 these are becoming so reduced in numbers by drought and air pollution that it would not be defensible to gather any but the commonest and most abundant for dyestuffs. They are used most extensively in northerly areas, where the high rainfall and lack of competition from other plants means that they are one of the predominant sorts of plant. In Scotland they were one of the chief sources of colouring for tartans and other traditional costumes. The brown dyes, or crottles, that are still used to give the characteristic browns to genuine Harris Tweed, come from a crusty rock-borne lichen called *Parmelia omphalodes*. Other species are used because of the dramatic colour changes that occur in them when they are treated with acids or alkalis. Corklit, or cudbear, once used extensively to dye soldiers' tunics, comes from *Lecanora tartarea*, a rather drab, grey species that grows on upland rocks. But extracted with ammonia it gives a vivid purple colour. In an altogether more earthy Highland formula, it was soaked in a mixture of stale urine and powdered chalk. The commercialisation of this process in Scotland in the 19th century led to one of the most bizarre examples of commercial refuse collection ever. Agents from the dye-factories toured the suburbs of Glasgow, and collected up to 3000 gallons of household urine a day. Each collector carried a hygrometer to prevent their being passed off with diluted samples.

Scents

We are probably living through a period where we are more conscious of scent than at any time since the 18th century. We have scented soaps and after-shaves, perfumed washing-up liquids and car-interior deodorisers, even fragrant lavatory paper. The latest development in perfume technology, micro-encapsulation, is an ironic echo of the older use of natural plant scents, for it almost exactly reproduces the method by which plants contain, and then release, their perfumes. Micro-encapsulation involves trapping minute particles of perfume essences in capsules no bigger than a grain of talcum powder. When these capsules are ruptured – by rubbing, say – they release their scent. This technological trick has led to the development of a rash of odorous gimmicks: scratch-and-sniff T-shirts, picture books whose illustrations give off a smell vaguely reminiscent of their subject when they are stroked, and even perfumed carpets. (See p. 140 for how they were making stamp-and-sniff carpets five hundred years ago!) Aromatic plants contain their scents structurally in precisely the same way, in a layer of oil glands on the outer surfaces of the leaves. Their function in the plant often seems to be as a guard against water loss, an oil-skin in reverse as it were. The oils evaporate in preference to the vital water supplies in the inner cells, and help keep the plant cool – which is why, presumably, so many of the herbs of hot, arid regions have aromatic leaves.

The new generation of synthetic household perfumes have none of the freshness and subtlety of the older natural scents, which were composed of a multitude of blended oils and resins. And such is their cloying proliferation that we may be approaching the decadent days of the Restoration, when the doyens of fashion paraded about with their hair caked with pomander, their bodies smothered in animal musks and bags of herbs strapped under their armpits to mask the odour of last week's stale perfume. The enjoyment of natural scents as a by-product of ordinary activities seems to me much more preferable. I can think of no neater or more satisfying functional use of a scent than the alfresco sauna baths which Sudanese women take in regions where water is too precious to use for bathing. They make a hole in the sand, and place in it an earthenware pot full of burning charcoal and branches from aromatic shrubs and herbs. They then take off their under-garments and crouch over the pot, drawing their cloaks around them to keep in the fumes. This promotes sweating, cleans the skin, and, as a bonus, leaves it with mild scent.

It's a custom which must be as enjoyable as it is effective; and there are many examples in this book of the functional use of plants being enlivened by their fragrance. (See BALSAMS, BIBLELEAVES, ELDER, FIREWOOD, FLEABANES, HEATHER BEDS, HOP PILLOWS, NUT POLISHES, RUSHLIGHTS, RUSHMATS, TOADSTOOLS, TOBACCOS, WICKER.) So I've confined myself here to a few notes about the general ways of using scented plants.

There is no shortage of raw materials. An enormous number of wild plants are scented, and often have quite different perfumes in their leaves, flowers and seeds. Just to run through some of their names – dame's violet, clove gillyflower, fragrant agrimony, musk rose, meadowsweet, angelica, sweetbriar, thyme, peppermint, sweet gale, balm, honeysuckle, spikenard, lily of the valley, sweet woodruff, juniper – will turn your mind to thoughts of walks through summer meadows, and remind you just how sensitive to scent were the unknown countrymen who gave our wild flowers their common names.

For the very best scented materials, pick your leaves or flowers in warm dry weather, preferably when the plant is just coming into full bloom. If you are going to use them dried, spread them thinly on a newspaper, and dry them – either in the sun, or in a warm room – until they are just brittle to the touch.

Two commercial ways of processing scented plants – distillation and enfleurage – are, I feel, too complicated for household use, and I would refer interested readers to the books on scent in the bibliography, where they are described in detail. But enfleurage is worth a passing note. This is a technique, used chiefly in France, in which successive layers of flowers are sandwiched between plates of glass coated with absorbent fats. The idea is that the fats will absorb just the volatile aromas of the plant, leaving behind the unwanted oils which are inevitably dissolved as well when scented plants are treated with liquid solvents. The aromatics are then extracted from the fats with alcohol. One endearing version of this process – as near a mechanical equivalent to a human sniff as I think it would be possible to devise – was patented by two British perfumiers in 1899. Reasoning from the fact that the scents of flowers are at their finest whilst the blooms are still unplucked, they developed an apparatus in which blocks of fat and oil-soaked cloths were suspended inside a greenhouse near growing flowers, whose aromas were gently wafted onto the absorbent substances by fans!

The easiest form in which to use scented plants is simply as they come, sewn into little muslin bags (which are sometimes known as sweet bags). Bags containing dried thyme and marjoram will help keep musty drawers fresh. I use one in the box in which I keep my loose copper coins. They can also be used to scent baths. Fresh sweet gale and sweetbriar give a warm, resinous fragrance to a bath (and a fine yellow colour) and pine needles or mint a cool, refreshing one. (Don't, by the way, use free-floating herbs in your bath or you will end up with more debris on your body than before you got in.)

Whole dried herbs are good to lay between freshly laundered sheets and clothing. Bunches of hay-scented sweet vernal grass and woodruff have been the most commonly used. You can make a pleasant little novelty out of the latter, for placing between clean handkerchiefs, by cutting the star-shaped leaf whorls from the stems (see p. 52).

Bunches of sweet vernal grass and woodruff, for scenting laundry.

If you wish to use scented herbs for shampoos or cosmetics you will need to make solutions of their aromatic principles. Most of these are soluble either in water or alcohol. A *decoction* is made by boiling the plant in water, but it isn't a technique I would recommend, as many of the volatile aromatic substances are lost by evaporation. A gentler technique is *infusion*, which involves steeping the leaves or flowers in warm water. The best way of doing this, especially with rose petals, is to place them in a closed glass jar with cold water, leave in the sun for a few days and then filter off the petals. If either the sunlight – or your patience – is lacking, you can produce a similar potion by steeping two tablespoons of fresh, chopped leaves or petals in a pint of hot water for about half an hour. Strain, and store in a well corked bottle. But remember that, even thoroughly sealed, these infusions tend to ferment after a while because of the interaction between bacteria in the air and plant sugars released by the hot water. *Tinctures* are made by steeping dried or crushed plants in alcohol. If you find biological alcohol difficult to obtain (chemists are sometimes wary of selling it) and can afford an exotic alternative, try vodka. Dilute the alcohol slightly (but not the vodka) in a clear glass bottle (it is worth being able to see the colours), add a large spoonful of the plant, cork tightly and leave for a week or so. There is, in this case, no need to strain off the plant material as you would with an infusion. Then use a few drops in your bath, or as a cologne or after-shave. Sharp-scented plants like mint, pine needles, fennel leaves and juniper berries are especially good for this. Similar lotions can be prepared by substituting white wine vinegar for alcohol, and can be used in rather larger quantities – in foot-baths for instance.

One of the best known uses of scented plants is in pot-pourris. These are really just mixtures of dried herbs and flowers left in jars to scent a room, and a good deal of unnecessary mystique and excessively extravagant recipes have gathered around them. The basic ingredients are dried flower petals, spices, and a preservative to absorb and fix the scents. Beyond that you can add anything you fancy. A simple pot-pourri of wild plants can be made by drying a few handfuls of wild rose petals for three or four days in an open bowl; mixing with some dried herbs (marjoram for sweetness, bog myrtle for spiciness and lemon balm for a citrus tang make a good blend); packing the mixture in layers in a glass jar with a close fitting lid, and sprinkling a little salt between each layer, which will help dry them further and hold their scent. This can then be used immediately as a pot-pourri. Better fixatives, which also add their own scents to the mixture, are the powdered root of orris (the southern European fleur-de-lys, *Iris florentina*) and dried and powdered oak-moss (the common lichen *Evernia prunastri*, which is still gathered in large quantities in France and Italy for use in the commercial perfume industry). You can later add spices, dried orange peel and fresh herbs and flowers to bolster up the scent.

If you do make pot-pourris, make sure you get the best out of them by using them properly. The jars in which they were stored were traditionally kept shut all day, and in the evening stood near the fire until they were warm. Then the lid was taken off and the perfume allowed to spread through the room. The hot jar was removed from the hearth onto a stand (which was often incorporated in the design of bespoke pot-pourri jars) until it was cool, and then shut again. Looked after in this way a jar will keep its fragrance for several years, and serve you better and more pleasurably than any commercial Airwick.

But the best rule to follow with pot-pourris – as with any scented-plant products – is to forget the recipes and follow your nose. Use the plants that please you most, and mix them as adventurously as if you were inventing a new cocktail. Carry your favourites about with you, stuck in the band of a summer hat, or just in your pocket, to sniff on a long car journey. Blend them into the wax if you make your own candles. Throw them on the fire, like those Sudanese women, or thread them between the slats of your sun-blinds. And don't forget the cat. A felt mouse, filled with dried cat-mint, *Nepeta cataria*, will prove an irresistible toy for most cats. The delight in natural scents goes right through the animal world, and there could be no better advice about how we could begin to revel in them more fully ourselves than in John Gerard's description of cats' behaviour in their own special mint:

Cats are very much delighted herewith; for the smell of it is so pleasant to them that they rub themselves upon it, and wallow and tumble in it, and also feed upon the branches and leaves very greedily.

The Herball, 1633

Catmint mouse.

Woods

If you are carving odd fragments of wood by hand (and there is nothing in this book for which you need more than a sharp penknife) your needs in terms of material are rather different from those of a commercial manufacturer. A mass producer of wooden furniture, for instance, needs a steady supply of timber of absolutely consistent qualities. You will be interested in precisely the opposite quality, *inconsistency*, in the way that shape and texture and grain pattern vary from one piece of wood to another. It is one of the excitements of carving wood by hand that you experience intimately just how much these qualities vary not just inside a single species of tree, but inside individual specimens. Unlike a potter or silversmith, say, who handles materials with standard properties, a woodworker knows that no two pieces of his stock-in-trade will ever be the same. Wood is the most complex of all raw materials. A single branch is, in its different parts, strong, whippy, dense and porous. It is surrounded by a soft outer covering that decays after a year or two's exposure to the weather, and a tough heartwood that will outlast most metals. Its grain pattern will be close and contorted where side branches emerge and straight and open between.

The secret of easy whittling is to look for pieces of waste wood that most closely correspond to the article you wish to carve. Take the making of an elderwood paper-knife as an example. You search for a length of fresh wood (about 2 cms thick) which has a long diagonal break. This will give you the blade already half-shaped. Elder is partly hollow (see p. 73), so by scraping away the pith and shaving the thin

Paper-knife whittled from broken elder branch.

outer wood you can quickly finish the blade. Peel off the stringy green bark, rub down with sandpaper, and you can produce a finished paperknife in about twenty minutes.

So in small-scale wood carving it is not so much a particular species you are after as a particular piece of wood. Nevertheless certain species do have qualities that make them especially suited for particular jobs. Soft woods are best to begin experimenting with, and there is none better than two or three years old elder. Lime is another soft timber that cuts very cleanly. The different effects that can be achieved with this cheesy wood are marvellously illustrated in the carved frieze by Grinling Gibbons (1648–1721) in St Paul's Church, Covent Garden. It is of a wreath of flowers, luxuriant and extravagant at the top, and as simple as a child's wooden toys at the bottom.

When mature, elder paradoxically also supplies one of the hardest woods, strong enough for making machine cogs. It has been used as a substitute for ivory, and, dyed black, for ebony. The best-known temperate zone hard timber, boxwood, is too scarce to be regarded as material for rough carving. But it was once in great demand for the blocks from which wood engravings were made, and Thomas Bewick claimed that one of his box-blocks was still sound after being used 900,000 times.

If you are after coloured woods, try some alder, which is easy to carve, resistant to damp, and has a fine, ruddy-brown tone. For small scale ornamental work, juniper is delightful, having a reddish bark, pink heartwood and white sapwood, like a stick of vegetable seaside rock.

Juniper wood is also finely scented, as are many other conifers. Cypress and thuja species are now being extensively grown in plantations for fence posts, as their aromatic resins make them resistant to attacks by insects and fungi. Once they were made into moth-proof boxes and chests.

For eccentric grain patterns there is nothing better than the bosses and burrs which grow on the trunks of some trees, particularly walnut and elm. Try also the whorled, red-tinged root stumps of hawthorns. Trees of the rose family – particularly pear and service trees – also have fine grain patterns, though these woods are hard to come by.

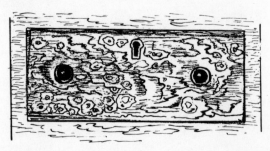

The beautifully convoluted grain pattern of walnut, *Juglans regia*.

If you want clean, durable and scrubbable wood, suitable for kitchen implements, say, try maple and sycamore. Sycamore is a pale, lustrous cream in colour and does not stain or darken appreciably even after long use. It has consequently been used for rolling pins, kitchen tables, bread boards and wooden plates. In dairy farming areas it has also been popular for making butter prints and spoons (see illustrations overleaf) – perhaps for the additional

PLATE 1. *Non-flowering plants*

1. Fly agaric, *Amanita muscaria:* pine and birch woods, August – November.

2. Shaggy ink-cap, *Coprinus comatus:* waste ground, roadsides, etc, May – November.

3. *Lecanora muralis*, (a *crustose* lichen, type b on page 113): rocks, walls, asbestos roofs, throughout the year.

4. *Parmelia saxatilis* (a *foliose* lichen, type c): trees and walls, throughout the year.

5. Oak moss, *Evernia prunastri* (a *fruticose* lichen, type d): deciduous trees, chiefly oak, throughout the year.

6. Razorstrop fungus, *Piptoporus betulinus:* birch trees, throughout the year.

7. Common horsetail, *Equisetum arvense:* green, non-reproductive shoots, April – October.

2

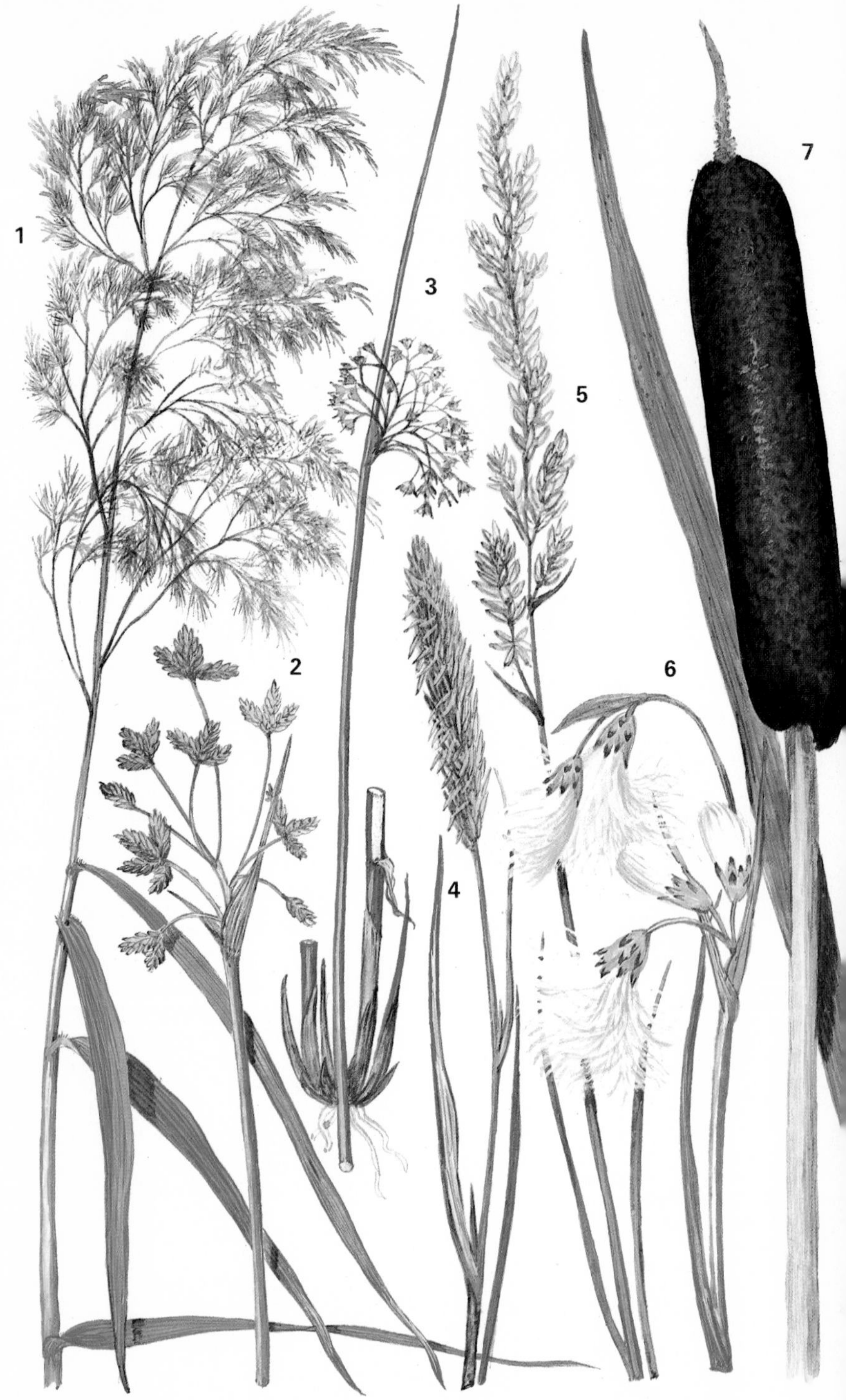

reason that sycamore trees are frequently planted as shelter for stock, and so are readily available as timber.

Beechwood has similar qualities, though it is less prone to warping or splitting if properly seasoned, and is the favoured wood for butchers' blocks. Incidentally, it is possible that the word beech and book have common origins, because the earliest 'buches' of northern Europe are reckoned to have been the smooth boles of beech trees on which the scribes carved their runes. Lovers still use them in the same way – not such a vandalistic practice as it is often made out to be, if done with care and imagination. It is a custom of very great antiquity, with even a Latin proverb to its credit: 'Crescent illae; crescetis amores;' – As these letters grow, so may our love.

Most of the timbers I have mentioned in the last few pages won't be readily available to the scavenger, and if you are simply looking for 'a bit of wood' to chip away at, the three species you are most likely to come across are 'deal' (pine, spruce, etc), birch, and elm. Commercial plantation conifer wood is not very satisfying to work with, but there is

PLATE 2. *Grasses* (Mostly harvested after flowering)

1. Reed, *Phragmites australis:* shallow fresh or brackish water, flowers August – November.

2. Bulrush, *Scirpus lacustris:* edges of rivers and lakes, June – July.

3. Soft rush, *Juncus effusus:* damp grassy places, June – August.

4. Marram grass, *Ammophila arenaria*, coastal sand-dunes, July – August.

5. Tussock sedge, *Carex paniculata:* marshes, fens, wet woods, May – June.

6. Common cottongrass, *Eriophorum angustifolium*, acid bogs, damp moorland, April – May.

7. Reedmace, *Typha latifolia*, swamps and the edge of fresh water, June – August.

A Welsh love spoon from the 19th century. These betrothal gifts, with their symbolically linked rings, were traditionally carved out of a single piece of sycamore.

always a good deal of it lying about on woodland floors. So is there often of birch, which the Scottish Highlanders exploited as an immensely versatile wood simply because there were very few other trees which grew there:

The Highlanders of Scotland make everything of it; they build their houses, make their beds and chairs, tables, dishes and spoons; construct their mills; make their carts, ploughs, harrows, gates and fences, and even manufacture ropes of it. The branches are employed as fuel in the distillation of whiskey, the spray is used for smoking hams and herrings, for which last purpose it is preferred to every other kind of wood. The bark is used for tanning leather, and sometimes, when dried and twisted into a rope, instead of candles. The spray is used for thatching houses; and, dried in summer, with the leaves on, makes a good bed when heath is scarce.

(J.C. Loudon, 1842)

There is also, of course, birch bark. Some American Indian tribes use this in seven different thicknesses, the finest for kindling fires, the thickest for making canoes.

Sadly, though, elm is now the most commonly available timber of all. H. L. Edlin has written a memorable passage on the traditional sources of this fine timber:

Sown by the wind, reaped by the whirlwind, elmwood may be ranked as a natural fruit of England's soil, and chance, unaided by man's planning, produced enough of it for the traditional woodworker's needs.

Woodland Crafts of Britain, David & Charles, 1973

As it turned out, 'man's planning' could not prevent

A traditional gipsy clothes-peg is easily made by shaving the bark off a piece of young willow about 1 cm wide; binding one end with a strip of tin cut from an old can with metal shears, which is then fixed with a short nail; and then cleaving the willow up as far as the binding and shaping the two prongs.

Dutch elm disease ravaging the elm populations of Europe and North America in the 1970's. There is such a glut of elm wood on the market that its selling price is less than the cost of felling and moving the dead trees. Although it is not the easiest of timbers to work, it is tough, resistant to splitting, very durable if kept damp, and with a fine 'partridge breast' grain pattern in the heartwood that makes it popular for floor boards and turning, particularly as it will polish up well. Elm disease has no effect whatsoever on the properties of the wood, and it is one small grain of comfort in an epidemic that will utterly change the face of the English landscape that an increasing effort is being made by individuals and local groups to find uses for a wood on which the timber industry is turning its back. Buckinghamshire County Council are making many of their new road signs from diseased elm. And with a proper sense of recycling, a nature reserve near my home has mounted its warning signs about diseased elms on poles made from their lopped branches.

This has been a short outline of some of the styles of use to which different kinds of timber can be put. For specific examples see the entries on ASHPLANTS, BROOMS, ELDER, FAGGOTS, FIREWOOD, FORKS IN TREES and SPINDLE.

Weaves

The finer ends of growing plants – twigs on trees, shoots on vines, rushes and reeds – have, because of their flexibility, been bunched, twisted and woven into a multitude of

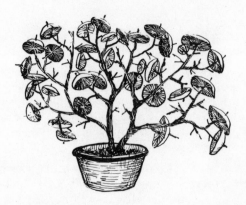

Thorns, at all times and in all civilisations, have been the universal vegetable awls, used for boring holes, tacking cloth together, tattooing skin, catching fish and enticing winkles from their shells. A pleasant and more practical home use is for drying mushrooms. Simply prop a thorny twig upright in a pot, and impale clean mushroom caps on the spines. Stand near a fire, or in some other warm, dry spot. The mushrooms are dry when they are brittle to touch, and can be reconstituted by soaking overnight in water.

A disposable rush spoon, Paiute Indian-style.

articles. There are some details of these in the sections on BINDWEEDS, BIRCH WHISKS, BROOMS, FLAXES AND FIBRES, HEATHER BEDS, REEDS AND RUSHES and WICKER. For one fine example, that perhaps sums up better than any other the principles of plantcraft that I have tried to outline in this book, let me mention the rush spoons made by the Paiute Indians. These were woven rapidly out of four or five green rushes, where and when they were needed, used for a few days until they began to dry out, and were then thrown away, to go back to the soil that produced them.

Ashplants and walking sticks

All work with plant material begins with an appreciation of the *growing* plant, an understanding of its qualities and potential as a living thing. This is especially true with trees. A craftsman in wood needs to know how a tree grows in response to light and wind, where its natural stresses are (and so how it will split), and be able to track down its 'muscles' from the way in which it performs the extraordinary feat of standing bolt upright for centuries. He also needs a sense of properness, of the right tree for the job. The most convincing work has always been produced by those who have seen this decision as a matter of belief and custom as well as mechanics.

The ash, the tree of modern pick handles and walking sticks, has a reputation for strength which goes back into prehistory. The Vikings made their axe handles from it. In Scandinavian mythology it was Yggdrasil, 'the greatest and best of all trees. Its branches spread over the whole world.' In the places where there were still primeval ash forests on the northern hills it must have seemed as if this were literally so. The ash can be an immense tree, growing up to 50 metres tall, straight-trunked, grey-barked, its twigs curving towards the sky. More than any other hardwood tree – more even, I think, than the ruggeder oak – it impresses with a sense of stateliness and restrained power. No wonder people hoped that, by using the tree, they would share a little in its strength. In the 16th century, ash keys were carried to ward off serpents and witches. In the remoter

Scottish Highlands, a spoonful of ash sap was the first nourishment to pass a newborn baby's lips. As late as the mid-18th century Gilbert White met villagers who had been through an ash ritual in childhood as a treatment for weak limbs or rupture. It was an extraordinary ceremony, a relic of pre-Christian sympathetic magic. A young ash was split and held open by wedges, while the afflicted child was passed, stark naked, through the gap. The split was then 'plastered with loam, and carefully swathed up. If the parts coalesced and soldered together . . . the party was cured; but, where the cleft continued to gape, the operation, it was supposed, would prove ineffectual' (*The Natural History of Selborne* 1789)

Belief in the power of the tree was one of the reasons behind the popularity of ash wood for spears and staffs – the strength of the tree lending itself, hopefully, to the user of the wood. But such beliefs rarely flew in the face of practicality, and I doubt if ash would have survived as the favourite wood for crooks and spears and walking sticks if it had not been physically right.

In fact ash is a strong timber, one of the toughest and most elastic there is. No other wood is so safely bent when seasoned, or is better able to withstand sudden shock and strain. At the time the trunks of those Hampshire ashes were being split to provide a magic portal for weak and ruptured children, their equally supple branches were being pollarded for wagon shafts and pitchfork handles. Today it is as much a sportsman's timber as a farmer's, used wherever there is a need for strength and shock absorption: oars, cricket stumps, billiard cues, hockey sticks – even police truncheons. There is a satisfying spring in the wood that you can feel just by pushing a branch against the ground. It will give a little when you rest your weight on it, a quality which is a considerable virtue in a walking companion!

The smoothness of the bark in the hand is a bonus. And so, for the producer of ash walking sticks, is the straightness and speed with which the young shoots grow. Most timber for the commercial manufacture of ashplants is specially grown. The sticks are either cut from two or three year old coppice (p. 77), or are grown to the required shape from seed. If straight poles are cut, the handles are formed by heating the sticks in damp sand and bending the softened wood in a curved vice. A more primitive method involves wedging one end between two rafters and hanging a heavy weight on the other.

The art of growing sticks with naturally curved handles is still practised in Surrey. Ash seedlings are grown in nurseries in the usual way, and are transplanted when they are one or two years old. But instead of being set upright, they are placed slanting in the ground with their end buds nipped off, so that the seedling has to use a side bud if it is to continue growing upwards. This new shoot – destined to become the shaft of the stick – rises almost at right angles to the old stem, which eventually becomes the handle.

A natural 'handle', formed by the upward growth in a well-laid hedge.

It's an ingenious method, but was suggested, I suspect, by nothing more complicated than the pattern of new growth inside a hedge. The craft of hedge laying involves a very similar technique. The saplings are partially sliced through, and bent over almost parallel to the ground, so that the new main shoots grow vertically upwards.

It would have been from such well-laid hedges – now sadly much rarer in the countryside – that most countrymen would have cut their sticks. There is a nice riddle that points up the opportunism of glimpsing a likely branch: 'When is

the best time to cut a walking stick?' 'When yer sees it!'
Which seems to me a splendid answer to all would-be
specialists. Making good use of a plant is less a question of
esoteric craft skills than of recognising its potential: 'when
yer sees it'. It needs an open mind as well as an open eye.
And in keeping your eye open for walking sticks, you will
begin to become alert to the shape of branches, to learn to
distinguish really dead wood from that which has been
recently cut, to choose one bark texture rather than
another, and see a finished stick, grain, knots and all, inside
a rather ragged branch.

But in the literal sense there is a 'best time' to go looking
for sticks. Most wood seasons better – and is more readily
available – in winter and early spring. This last point, about
availability, isn't a trivial one. Since it is not advisable to go
cutting timber from other people's trees, you will need to
concentrate on areas where wood is cut seasonally anyway.
Most hedges, for instance, are trimmed back between
December and March. These days most of the cut wood is
either burnt on the spot or left to rot – or to be gleaned. Ash
cuttings are amongst the best to find at this time of the year,
and not just because of sticks that can be made from them.
They will be spiked with sooty black buds and perhaps the
first flowers, which Geoffrey Grigson described as 'little
crisp bushes like the carrageen seaweed in miniature; or
like hydroids from a sea pool, enlarged'.

But don't limit yourself to ash. Almost any sturdy timber
will do for a walking stick, if it takes your fancy. Sweet
chestnut is straight and damp resistant. Cherry has in-
triguing grain patterns, a colour close to a pale mahogany,
and a slight but pleasant fragrance. Hollies are rarely cut
back, but the wood, if you should find some, is hard and
dead white.

You need do no more to a stick than trim it to length and
make sure that it is comfortable to hold. But if you want
something more finished, you could season it for a few
months, peel off the bark, smooth any knots and burrs
down (but not away) with a surform, and then shine it up
with linseed oil or a clear varnish. Embellishing the stick
with carving is altogether more elaborate, but the cutting of

rings in the bark of staves is one of the oldest, and simplest, decorative arts in the world. It is particularly striking with hazel sticks (p. 77), because of the contrast between the dark brown bark and the white wood, and the suggestion of a flecking, a notching, in the natural pattern of the bark.

The lower branches of the blackthorn, too, make hard and finely coloured sticks, with a very aristocratic knobbliness. William Cobbett has some typically double-edged praise for these, in an essay on the delights of a surrogate port made from the fruits of blackthorn (sloes) soaked in brandy:

This Thorn . . . sends up straighter shoots from the stem than the White Thorn does, and these shoots send out, from their very bottom, numerous and vigorous side-shoots, all armed with sharp thorns. The knots produced by these side-shoots are so thickly set, that, when the shoot is cut, whether it be little or big, it makes the most beautiful of all walking or riding sticks. The bark, which is precisely of the colour of the Horse Chestnut fruit and as smooth and as bright, needs no polish; and, ornamented by the numerous knots, the stick is the very prettiest that can be conceived. Little do the bucks, when they are drinking Port wine (good old rough Port) imagine that, by possibility . . . for the 'fine old Port,' which has caused them so much pleasure, they are indebted to the very stick with which they are caressing their admired Wellington boots.

The Woodlands, 1825

(The most notorious blackthorn stick is, of course, the shillelagh, definitively described by the Chairman of the Pharmacology Department at the University College of Los Angeles as 'an ancient Hibernian tranquilliser'.)

The oddest sticks are those that have been naturally sculptured from a growing branch by honeysuckle. Honeysuckle is a woody climber with an iron grip, and as it spirals clockwise round a supporting tree, it constricts the soft outer tissues of the growing wood. The result can be a piece of wood as eccentric as a corkscrew. Such 'barley-sugar' sticks, as they were called, were in fashion not so long ago with music hall comedians. But previously they were part of the accoutrement of forest workers, a sign of their privileged relationship with wood and trees. This is how

A 'barley sugar' stick, formed by the pressure of honeysuckle on a growing sapling.

Thomas Hardy described the stick-carriers gathered round a Dorset auctioneer:

His companions were timber-dealers, yeomen, farmers, villagers, and others; mostly woodland men, who on that account could afford to be curious in their walking sticks, which consequently exhibited various monstrosities of vegetation, the chief being corkscrew shapes in black and white thorn, brought to that pattern by the slow torture of an encircling woodbine during their growth, as the Chinese have been said to mould human beings into grotesque toys by continued compression in infancy.

The Woodlanders, 1887

Balsams

A friend of mine has a balsam poplar in her garden. It's an awkward looking tree, pollarded too high up so that it reminds you of a long log, perched on end. But every April, as the leaves unfurl through the resin-coated bracts and sail stickily in the spring easterlies, a glorious incense fills the lanes round her cottage. In her second spring there, a neighbour called 'to thank her for the pleasure her tree had given the village'. It is not so often that you can, without exaggeration or licence, give thanks for a balmy breeze.

Balm, balsam, basme – they all derive from the ancient Hebrew roots *bot smin*, 'the chief of oils', and *besem*, 'a sweet smell'. The original balsam, Balm of Gilead, was the resinous sap of the Middle Eastern tree, *Commiphora opobalsamum*. It was used as an antiseptic and as a fragrant and soothing cosmetic, and gave us the second meaning of balm as a healing influence, a consolation.

Balm of Gilead was always a rare and jealously guarded tree, and most balsams are now extracted from commoner but similarly scented firs and poplars. In the case of the North American balsam poplars, it is not the sap that is used but the resinous leaf buds. The aromatic oil is separated from the buds by boiling them in water, and is exported to Europe under the name of 'Tacamahac'. It has been used in ointments as an antiseptic, and as an ingredient in tinctures for coughs and clogged lungs. The buds were listed in the United States National Formulary as recently as 1965 – one of the many examples of the official recognition of native folk-remedies. In this instance the medical establishment borrowed their remedy for winter catarrh from the Menominee and Ojibwa Indians, who used to warm the buds in fat and plug their nostrils with the resulting salve. You may do the same, though it is easier and rather less messy to throw the buds into a bowl of hot water and inhale the vapour. It is a warm and spicy smell, not at all strong, but wonderfully refreshing if you have a head cold. And through the scented steam you can watch the young leaves open, like jasmine flowers in a teapot.

The poplars are a vast and confusing tribe, with names that seem to hybridise as readily as the trees, and smell is probably the easiest way of picking out the balsams. A good sized western balsam poplar (*Populus trichocarpa*) can be nosed out from over 100 metres on a warm spring day. The heart-shaped leaves open early in April, only a short while after the long yellow catkins. They are large, heavy, a shiny bright green on top, but washed with white below. In winter the tree is just a characterless skeleton: grey-barked, weak-limbed, unimaginative and unambitious in shape, an outsize sapling that never grew up. But this is the season when the pointed, chestnut-brown buds form. They are already scented in October and become more so as spring approaches. By February or March they are sticky with yellow balsam, and this is the time to pick them. (Though don't forget, in May and June, that with a few of the leafy twigs you can scent *and* decorate a room.)

The western balsam is probably the commonest species, native across much of North America (where it is often known as the black cottonwood) and widely planted in plantations and along roadsides in Europe. It is also the tallest broadleafed tree in either area, sometimes reaching over 60 metres in height. The eastern balsam poplar (variously known as *P. balsamifera* or *P. tacamahaca*) is a more modest tree, with smaller leaves and a smoother trunk. The Balm of Gilead poplar (*P.* × *canadicans*) is a hybrid, possibly between the last tree and the American cottonwood (*P. deltoides*). As one writer put it, the genealogy of the poplars 'is like one of those exhausting novels which describe the adventures and alliances of several generations of an international family established in different capitals'.

Balsamic resins for coughs and colds can be obtained from other trees. Benzoin, the basic ingredient of 'Friar's Balsam', is the sap of far Eastern trees of the *Styrax* family; Balsam of Tolu, of the South American tree *Myroxylon balsamum*. Resinous secretions of this kind often seem to play a protective role in trees, and this is certainly true of pine resins, the most accessible alternatives to these tropical balsams. All parts of a conifer tree carry resin, which consists of two main elements: a volatile oil (turpentine), and a solid (rosin). If the bark is cracked, or a branch broken off, resin quickly oozes out to block the wound. After a while the turpentine evaporates from the outer surface and the rosin forms a dry scab. This seal provides an effective barrier to the spores of harmful fungi which might otherwise be able to take a hold in the moist and damaged tissue. But it also looks as if resin may possess some *chemical* antibiotic properties which inhibit the development of fungi and bacteria – a clue, perhaps, to their antiseptic action in human wounds.

Douglas firs. The resin which seeps from the bark (and forms blisters in crevices round the trunk) is known as Canada turpentine, and in solution is widely used as a mounting fluid in microscopy.

Apart from the small quantities used in medicine, pine resin forms an important ingredient in paints and varnishes, and is gathered by tapping suitable trees. Small nuggets for the inhaling bowl are best collected from cavities in the smooth trunks of Douglas and silver firs, where it forms blisters that can be prised out with a penknife.

Bamboo pipes

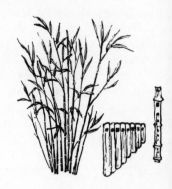

In mythology, Pan made his first pipe out of a reed, cut from the bed in which his beloved nymph Syrinx disappeared. It *is* possible to make a rough and ready whistle out of a reed or other hollow grass, but Pan must have had superhuman wind to have persuaded music from such narrow tubes.

PLATE 3.   *Scented plants*

1. Scots pine, *Pinus sylvestris:* forests, heaths. Scented parts: needles, resin.

2. Bog myrtle, or sweet gale, *Myrica gale:* wet heaths, bogs. Twigs, leaves, cones.

3. Balsam poplar (a narrow-leaved variety), *Populus gileadensis:* plantations, hedgerows, etc. Leaf buds and leaves.

4. Marjoram, *Origanum vulgare:* Calcareous grassland and scrub. Leaves and flowers.

5. Common valerian, *Valeriana officinalis:* woods, grassy places. Jasmine-scented flowers, June – August.

4

3

5

It is easier and more effective to use the wider and naturally hollow stems of shrubs like elder and bamboo. Bamboos are mostly natives of the far East, but are being increasingly found as naturalised escapes in Europe and North America. Clumps are very distinctive, with their tall sheaves of canes, and dense, pale green foliage. To make pipes, season the canes (or elder stems) well, and then hollow out a selected length with a hot metal rod. Shape a cork to fit one end, but before gluing it into place cut a segment away to make it flat on one side, to form the hole through which you will blow. Cut a small square hole in the top of the pipe, far enough away from the mouthpiece to be clear of the cork and your lips. Bevel the edge furthest from the mouthpiece, blowing the pipe all the while you are cutting to make sure the tone is right. To make what we understand as 'Pan pipes' today, prepare a set of these pipes of different lengths corresponding to the notes in a scale, and glue them together in a row. Alternatively you can make a recorder, by drilling or burning a set of holes along the length of one cane. Start with a very small hole a short distance up the cane from the square hole, which will form

PLATE 4. *Scented plants*

1. Sweet briar, *Rosa rubiginosa:* Calcareous scrub. Scented parts: apple-scented leaves.

2. Meadowsweet, *Filipendula ulmaria:* wet meadows, fens, ditches, etc. Leaves and flowers.

3. Pineapple weed, *Matricaria matricarioides:* waste places, arable fields, footpaths. Pineapple-scented leaves.

4. Salad burnet, *Sanguisorba minor:* dry calcareous grassland. Cucumber-scented leaves.

5. Balm, *Melissa officinalis:* naturalised on waysides, etc. Lemon-scented leaves.

6. Catmint or catnip, *Nepeta cataria:* hedgebanks, waysides. Leaves.

the 'ray' note in the scale. Tune it by enlarging the hole with a round file, and repeat this with six more holes up the cane.

A bamboo recorder.

The wood of bamboo is dense and resilient, and being hollow, is resonant when struck. It has been suggested that the system of pivoted bamboo pipes used to irrigate rice fields in China, and which strike sonorously against each other when they are shifted by wind or hand, originated the idea of several simple instruments: xylophones, for instance, made of sets of hollowed canes; narrow drums; and a kind of zither fashioned from a piece of bamboo, with strings slit from its own outer bark and held away with tiny bridges.

Bibleleaves and Bookmarks

In the Dorset village of Tolpuddle there is a much-loved sycamore known as the Martyr's Tree. It is well over two hundred years old, with a girth of 6 metres, and since the late 1960's has been preserved with the help of a grant from the Trades Union Movement. Yet, to tell the truth, it is not much to look at. It was ruthlessly pollarded in the last century, and much of the heartwood has rotted away. The hollow trunk has been stuffed with vermiculite, and bound with iron hoops to prevent it falling apart. The reason for the care and affection lavished on it has little to do with its looks. In the 1830s Tolpuddle was the birthplace of what was probably the first agricultural trade union; and it was under the village sycamore that the local farmworkers used to meet and talk, an illicit custom then that ended in their transportation to a penal colony in Australia. They became known as the Tolpuddle Martyrs, and the tree under which they met became a symbol both of their own association and their association with the life of the earth.

I have touched on this background partly because there is a story that the Martyrs' spokesman, George Loveless, took a leaf of the sycamore to Australia with him, pressed between the pages of his Bible; and partly to underline the complexity of our loyalties to plants. If an ungainly leaf from a southern European immigrant seems an odd memento of an English village, it is no more odd than the preservation of a rotting specimen of what has become virtually a tree weed. Our feelings about plants are, thank goodness, rarely so rational or considered. Plants have meanings for us, and if we lose the plant, or leave it behind, we lose part of the meaning.

So we put plants in books as tokens and keepsakes, mementoes of walks and places and flowers picked by loved ones – or simply to remind us to look up their names. There they usually stay, flattened against a poem or their own portrait, until they crumble into dust. That is their place, where they were meant to leave their mark.

But some leaves, because of their scent or their durability, have been used as bookmarks in a more general way. There are a scattering of Bibleleaves and Book-leaves amongst the lists of country names for plants. Tutsan, *Hypericum androsaemum*, has been one of the most popular, as much, perhaps, because of its reputation as a beneficent herb as for the shape and fragrance of its leaves. (Medieval herbalists mistook it for the *Agnus Castus* of Pliny which 'drywyth awey the fowle lust of lecherye if men drynke it'.) But tutsan's leaves are virtuous enough at face value. They are heart-shaped, up to 10 cm long, and develop a sharp, sweet fragrance when they dry, a tang like fresh pipe tobacco or Christmas cake.

Tutsan is a shrubby, more or less evergreen member of the St John's wort family. It is not uncommon in damp woods and shady places, especially along the steep hedge banks in Devon and Cornwall. The bushes are covered with

The broad leaves of costmary (above), and tutsan, both used as bookmarks.

bright yellow flowers in June, which give way to shiny berries, green at first, then red and finally purplish-black.

If you pick and press a leaf in summer, it will begin to smell within a few days, and keep its aroma for up to four years. But don't expect the scent to waft out of the book as you open the marked pages. Bookleaves need an appreciative sniff – and, for that matter, a warm, damp day to bring out the best in them.

Another plant which develops its scent on drying and which is architecturally ideal as a bookleaf, is sweet woodruff. In this case a sprig of the whole plant is used. Woodruff grows to about 40 cm in height, and carries up its stem rosettes of bright green, pointed leaves. With a little care you can tease these out flat on a page so that they keep their star shape intact. Press them immediately, and they will hold it for years. After only a few hours the woodruff will start to develop its perfume, a faint scent of new-mown hay at first, but strongly tinged with almonds after a few days.

Woodruff grows in patches on woodland floors and under dark hedgerows. It comes into flower between April and June and is such a welcome sign of spring with its shiny white flowers and compact leaf ruffs that I always feel uncomfortable picking even small bunches to flavour spring drinks. But one spray for a bookmark will trouble no one's conscience, and certainly not the woodruff, which, being a perennial, will soon make good its loss.

Costmary, *Balsamita major* – variously known as alecost, mint geranium and herb Mary – is, strictly, a native of the Orient. But the size and aroma of its leaves have made it a favourite bookleaf in Europe. The scent, a cross between mint and lemon, is pleasantly obvious even in the youngest leaves, and is the reason the plant was brought into herb gardens in the West. (It had a multitude of domestic uses, from cooking and brewing to the making of pot-pourris.) It has become naturalised in southern Europe and America, and still hangs on in a few old gardens in Britain, though it rarely flowers except in very warm summers. The leaves aren't unlike small dock leaves in shape, and can grow up to 20 cm long.

I should add that provided they are picked in an un-damaged state, and are well dried, none of these leaves will stain the pages of a book.

Bindweeds

Bindweeds must be amongst the gardener's most hated enemies. Invective and weedkiller are poured upon them – usually with absolutely no effect. The greater bindweed, *Calystegia sylvatica*, the large white-flowered variety of gardens and waste places, has a battery of survival tactics that is more than a match for our technology. The roots, for a start, lie so deep that they are beyond the reach of most herbicides. If they are pulled up, the plant can grow again from the smallest overlooked fragment of root or stem. If they are simply damaged they exude a milky sap which quickly congeals and seals the wound. The plant grows so fast that it can throw an entire loop around a stem or fence in less than two hours; and so extensively that it can reach the top of a sizeable tree. If it has nothing so substantial to support it, it will simply twine round itself. No wonder that amongst the host of names it has been given as a tribute to its snake-like tenacity – bellbind, withywind, strangleweed – it is also more universally known as Devil's guts.

But if you can't beat it, why not use it, and use particularly its remarkable qualities of strength and sinuousness? I have found bindweed ideal as a short-term, makeshift garden twine (a use that has some poetic justice, I feel!). It grows in considerable lengths, and is the only climbing plant I know which will take a tight reef knot without breaking. Bindweed also constructs its own ropes when it is forced to twine around itself, and these will do for heavier jobs.

Bramble is the plant which has been used most frequently as a natural twine, especially for broom and basket making. If you pick up a dead bramble on a road or path, after it has been well-pounded by tyres and feet, you will see the long fibres that it needs as a clamberer, and which give it its strength as a twine. Woodland brambles are preferred, as they normally have fewer side shoots. The selected shoots are dampened, and the outer skin and prickles scraped off. The bramble is then split into halves, or quarters, by placing a knife against the thinner end and drawing it towards the thicker. The pith is scraped out, and the bramble finally shaped to the desired width.

Many other plants have been used as tying materials, including most of those used for making baskets (see p. 161). Barks and basts (stringy inner barks) are popular, particularly those of ash, hazel, lime, oak and sweet chestnut. Ash bark is removed from young logs after they've been soaked in water. A straight log about two metres long is stripped of its waterproof outer bark, and then laid in a stream (though a pond or waterbutt will do). After a couple of months it is taken out and rested across some sturdy trestles or chopping blocks to dry. As it dries out, the rings of wood representing one year's growth separate and split into strips the same length as the log and the thickness of veneer. They are helped on their way with a few light blows from a wooden mallet. Further taps will split the lengths still narrower, until they are the right width for binding or weaving, and also help make the wood fibres more supple.

Also used have been the thin shoots of willow, honeysuckle and dogwood, and the exceptionally pliant, almost unsnappable young twigs of the wayfaring tree ('hoarwithies'). Nettle fibres have even been spun into ropes, like those from its cousin, hemp. But none of them can take a reef knot like a common garden bindweed.

Birch Whisks

Of all the lovely parts of this loveliest of trees – the wisps of white bark, the pale gold autumn leaves – it is the birch's twigs which contribute most to its sense of grace and elegance. They cascade from the branches, as fine and intricate as lacework. So delicately are they hung that they quiver like tassels in even the lightest breeze. But in stronger winds the twigs start to lash about, and foresters will often remove colonising birches from mixed plantations to prevent them flaying the other saplings.

All 60-odd species have these thin and supple twigs, and throughout the northern hemisphere they have been used wherever there is a need for sweeping or whipping. A kitchen whisk is, in design at least, the prototype of all birch-twig implements, and is the easiest to make. Cut a handful of twigs about 40 cm long from the very tips of the branches, preferably in the winter when they aren't 'green'. Choose between thirty and fifty of the straightest, trim off any burrs or side shoots, and cut them all to the same length,

A whisk made from birch twigs – ideal for beating eggs, sauces, cream, etc., especially as it will bend round to the contours of the pan.

20 to 30 cm, according to the type of whisk you want. You can also, if you want a cleaner finish, scrape the bark from the twigs. But this isn't essential, and the bark should not start to crumble even during quite vigorous whisking. Then simply bind the twigs together at their broad end with a length of string or wire. In Sweden they sometimes use another fine birch twig as the binder, splitting and moistening it so that it will automatically tighten up on itself as it dries. But this needs a considerable knack, and it would be sensible to practise first on something rather more solid and stable than a handful of twigs.

Birch whisks are very satisfying to use (though rather fiddlesome to wash); and, as one cook remarked to me, are so good to *listen* to.

Brooms and besoms

A birch broom, or besom, is really just a large whisk, and at a pinch you can make one in the same way, by binding together a bundle of long twigs. But why is such an implement called after another, unrelated, species of shrub? An only marginally slimmer switch is, after all, canonised as *the* birch! But it seems likely that brooms, for sweeping, were made from broom, *Cytisus scoparius*, some time before the less obvious twigs of birch were used, and gave their name to the universal sweeping instrument. (The history of the word doesn't give us much clue, the old English root, *bröm*, meaning nothing more specific than 'a thorny shrub'.)

Broom – known locally as basom and brushes – is certainly an ideal plant to sweep with. Its young branches are long, straight, slender, springy, often leafless, and grow in such tightly packed bunches that it is possible to make a rough and ready brush by cutting a spray and using it as it comes. The first I made myself was from a spray I found on a derelict railway line in Norfolk, where they had been cutting back the broom to keep the footpath open. It was June, and all along the sandy track the bushes were blazing with the vanilla scented flowers. But this branch, cut conveniently with a long handle of older wood, had dried off in the summer sun. All it needed to make it functional was a few loops of string to bring the shoots closer together.

A switch of broom. Branches – and sometimes whole bushes – of broom's spiny cousin, gorse (or furze, *Ulex europaeus*), were once used as chimney brushes.

Broom in displays as magnificent as I saw that day must once have been much commoner. It is a plant which revels in poor, lime-free soils, and is one of the first shrubs to colonise sand dunes, heaths and open ground in conifer woods. It is common throughout temperate Europe and naturalised in a few sandy places in north America. The large, golden pea-flowers which cover the bushes between late April and June, make broom one of the great landscape plants, and architects often deliberately plant it out on the verges of major roads. But, as with so many 'economic' species, broom's most conspicuous identifying feature is the one that has subsequently had a use found for it – in this case the bunches of whippy green twigs.

C. scoparius was so common once, so accessible, and so right for the job that it can only have gone out of fashion as the basic stuff of brooms because it was vanishing from the landscape as heaths and commons were taken into cultivation. (Though in Scotland, where there is still plenty of it on the drier moors, its traditional use persists in the brushes used to sweep the ice in the ancient game of curling.)

Incidentally, the popular association of broomsticks with witches isn't all whimsy. Broom was looked on as a magical plant as well as a practical one. It was used by witches, and against them. Sprigs of the flowering branches were worn to ward off elfin lovers – and more earthly seducers, for that matter. In the old Ballad of Broomfield Hill, a lady has an assignation with an energetic knight amongst the broom, and fears she will lose her maidenhead. But a witch assures her the knight will fall asleep, and advises her of a way to ensure he remains so:

> Ye'll pull the bloom frae off the broom,
> Strew't at his head and feet,
> And aye the thicker that ye strew
> The sounder he will sleep.

These days broom's efficacy in troubles of the heart is more literal, and it is of accepted value as a mild cardiac stimulant.

Heather or ling (see p. 98) is another shrub whose form, abundance and handiness suggest that it must have been picked out early as a sweeping plant. In moorland areas, heather becomes the most usual material for brooms, and even appropriates some of the local names for birch and broom – e.g. bissom and bazzom.

Heather for brooms is cut in the spring, as soon as the twigs have become green and pliant. If it is cut before this, in its sapless, 'sleeping', winter state, it is apt to shed dead leaves and bark dust as it sweeps, and deposit more on the floor than it takes off. Ideally, broom ling should be tall and straight-growing, with long stems and a fair amount of young branching. A bundle of sprays is clamped in a vice (the professionals use one with semi-circular jaws), and then bound up with ash strips or, these days, with wire.

The ends are then squared off with a chopper. The handle ('tail') can be made out of any straight pole, though ash, hazel and beech are the favourite timbers. It is fitted by being sharpened at one end and pushed into the broom head, and driven home by banging its free end against the ground. When the pole has penetrated a sufficient distance to hold firm, it is secured by a nail driven through both head and tail.

Marram brush.

The making of birch besoms is fundamentally the same, though in this case the twigs are cut in winter, and seasoned for a few months. (In the Bridewell Museum in Norwich, there is a glorious photograph of a bewhiskered broom squire, holding before him a new-made besom still decked out with its yellow March catkins, like a Palm Sunday garland.) Commercially, birch for brushes is grown in coppice (see p. 77) and is cut from the crown of the trees when they are at least seven years old, though for home manufacture any birch brushwood will do. After seasoning, the twigs are trimmed to about metre lengths, and the rough thicker twigs sorted from the fine. The rougher twigs are gathered into a core, and the finer ones bunched around them. The whole bundle is then tightly bound with wire. Broom-makers use a contraption called a horse for this. The operator sits at one end of a frame, whilst the other carries a foot-operated clamp to hold the wire as it is being tightened round the twigs. Finally a tail is driven into the heart of the brush head in exactly the manner described for heather brooms, which incidentally packs the twigs still more tightly against the wire.

Birch besoms have accumulated some curious industrial uses. The heads are used to line the vats in vinegar refineries, where they help to clarify the liquor, and they're still employed in the steel industry for sweeping away impurities from freshly pressed plates. Wooden brooms are preferred as they burn away on the hot steel, and take up impurities chemically as well as sweeping them off mechanically. But it is for the rather more gentle business of sweeping leaves from lawns that besoms are best known and most widely bought. There is nothing else which can so effectively ambush leaves on a spongy surface than this tangle of springy twigs. A specially designed synthetic material for sweeping leaves from grass would, I think, be forced to imitate the birch twig's natural bounciness and rough, irregular surfaces.

A variation on the bundle of twigs formula is found in Anglesey, where they make hand brushes and brooms out of the tough marram grass (see p. 130) which grows abundantly on the sand dunes. In the simplest form, a rectangular hole is cut or burnt at the end of a slat of wood, and a bundle of grass threaded through from each side, so that there is an equal number of stumps and tips on both sides of the slat. The grass is then tied beyond the end of the slat, which fans it out and produces a very serviceable brush. The making of full-size brooms is a more elaborate business. Two long bundles of grass are bound to the end of a tail, one with its broad ends pointing away from the handle, the other, longer bundle is fixed in the opposite direction so that its stumps slightly overlap the tips of the first layer and its own tips point up towards the handle. This outer sheaf is then turned down over its binding, so that its tips cover the stumps of the first bundle. As each blade is turned down, a binding cord is woven alternately above and beneath it. It is

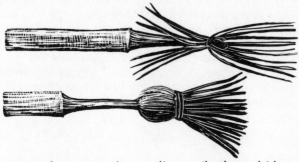

The stages in constructing a splint, or 'backwoods' broom.

a difficult process to describe in words, but the individual binding of the strands produces a broom that is as firm as it is attractive.

This 'bending-back' principle can be seen at its most ingenious in the 'splint' brooms of central North America. (I've also seen them referred to as 'backwoods' brooms, which would be a more accurate description, if one could believe it was a pun!) They have the distinction of being the only brushes made out of a single piece of wood. A straight pole at least a metre long and about 8 cm in diameter is debarked and partially shaved at one end. The shavings need to be long (up to 50 cm), thin enough to bend but not break, and to stop at least 3 cm from the end of the pole. As they are cut, the strips are carefully bent over away from the handle, and when enough have accumulated, are bunched and bound. Finally, the remainder of the now wasp-waisted pole is pared down to produce a handle of an even width. The timber for these brooms needs to be strong yet elastic, otherwise the shavings will snap as you bend them over. Hickory is most commonly used, then ash and oak and occasionally birch.

Besoms have a comfortable and rather bluff feel about them. Yet there are plant brushes of an almost mandarin delicacy – and, at the other extreme, a fearsome herbal scourge for improving the circulation. It was Gilbert White again who described how the dried fronds of the moss *Polytrichum commune* were used in much the same way as a feather duster:

While on the subject of rural economy, it may not be improper to mention a pretty implement of housewifery that we have seen nowhere else; that is, little neat besoms which our foresters make from the stalk of the *Polytricum commune*, or great golden maiden-hair, which they call silkwood, and find plenty in the bogs. When this moss is well combed and dressed, and divested of its outer skin, it becomes of a beautiful bright chestnut colour; and, being soft and pliant, is very proper for the dusting of beds, curtains, carpets, hangings, etc.

The Natural History of Selborne, 1789

Polytrichum commune, now usually known as hair moss, is a tall and distinctive species of damp, acid soils, especially on wet moorland. The stalks can grow up to 20 cm tall, and are covered with fine, slightly toothed leaves.

It would be difficult to think of a more contrasting plant to hair moss than butcher's broom – known more graphically as kneeholm and knee holly. It is one of the most curious of all woodland plants: a stiff, spiny shrub, growing knee-high, with prickly leaves that are not leaves at all but flattened stems, and which bear in their centres first the tiny greenish-white flowers and then the scarlet berries. To complete the confusion it is not a holly, or any kind of woody shrub for that matter, but a member of the lily family.

The use of the dried twigs by butchers for scouring blocks and benches goes back at least as far as the 16th century, when the name butcher's broom first appears in herbals. John Parkinson, writing in 1640, says that 'a bundle

Butcher's broom.

of stalkes tied together serveth them to cleanse their stalls, and from thence we have our English name of Butcher's broom'. But he also mentions another butcher's gadget, a miniature indoor hedge made of the thorny twigs and hung round meat to keep off the mice. It is debatable which of these came first, and perhaps inspired the other. By the 19th century both functions had probably fallen out of fashion. But old practices die hard, and tend to live on in customs long after their original purpose has been forgot-

ten. W. A. Bromfield, writing in 1856, described how the
butchers of his time decorated 'their mighty Christmas
sirloins with the berry-bearing twigs'. An echo, perhaps, of
those lean-to meat safes?

The last recorded commercial use I can trace also had
tenuous connexions with the meat trade. Apparently there
was nothing to equal the sharp-pointed leaf-branches for
sprinkling water evenly over raw chamois leather and
parchment, and the plant was cultivated for the leather
industry at Whittlesford in Cambridgeshire.

Throughout all this ingenious exploitation, butcher's
broom was, as you would expect of so singular a plant, in
great demand as a herbal remedy. Culpeper, always a great
believer in nasty medicine, recommended a poultice of the
leaves for mending broken bones. But not even he had the
stomach to pass on its ancient employment in the flogging
of chilblains.

Butterbur

Would it occur to us, I wonder, to use the huge floppy
leaves of butterbur, *Petasites hybridus*, for wrapping *butter?*
I doubt it. They might suggest a tablecloth, or even a
makeshift umbrella. John Gerard reckoned the leaves to be
'of such widenesse, as that of itself it is bigge and large
inough to keepe a mans head from raine, and from the heate
of the sunne.' But we are too accustomed to refrigeration
and airtight shrink-wraps to think that a vulnerable food-
stuff might once have had need of a bit of natural shade.

Before the days of cold storage, it must have been imperative to find a material to keep butter cool and contained. Butterbur leaves, big and shady enough for chickens to shelter under in the heat of the summer, must have turned men's minds to cooler things just by the looking. They felt right, too. They were pliable, non-conductive, thick enough to cushion the butter from bruising and to soak up any seepings and meltings.

Butterbur leaves start to crowd over damp waysides and stream banks rather late in the summer. Like its relative, coltsfoot, its flowers appear first, often as early as February. When they first push through the soil they look sufficiently like slightly flushed button mushrooms to have earned the country name of 'early mushroom'. But later the tasseled flowers make butterbur look more like some dwarf pink conifer. (These flowers have their uses too, being an invaluable source of nectar at this early season. In Scandinavia butterbur is often deliberately planted round beehives.)

Butterbur's original use is redundant now, but try it as a makeshift wrapping material when you're out-of-doors – for your head, as Gerard recommends, for picnic left-overs, or for a trout from a stream, if you're lucky enough. It is a delight to handle. The soft grey down on the undersides of the leaves is invitingly cool, and they fold up as neatly as vine leaves in *dolmades*.

Although they can't really match the pleasing, lint-like furriness of butterbur, a number of other leaves have been used as butter-wraps, notably burdock and the larger-leaved docks. And the boldly-veined leaves of one American plant have been used not just for wrapping, but for pressing onto and patterning butter and pastry (much in the manner of wooden butter-prints). This is the Indian mallow or velvet leaf, *Abutilon theophrasti*, a native of southern Asia which is now well-established as a weed in the warmer parts of the United States, and cropping up increasingly frequently in waste places in Europe. Velvet leaf is an annual about a metre tall, unbranched, hairy, and carrying small yellow or orange flowers. The veined leaves are heart-shaped and woolly, and their domestic use has given the plant local names such as butter-print and pie-marker.

Velvet leaf, *Abutilon theophrasti.*

I have come across one similar practice in England, which involves the use of sycamore leaves. In the West country they make small rich Easter cakes called Revel Buns, and each bun is baked on its own individual leaf. 'I do not know why they must be baked on sycamore leaves' remarked one Devon lady, 'but as a child I should not have thought them right if they had not the imprint of the leaf under them'. (Mentioned in *Good Things in England*, ed. Florence White, 1932).

Daisy chains and dandelion clocks

I hope that future generations of plant economists don't follow their zoological colleagues, and interpret children's plant games as nothing more than a rehearsal for more serious and utilitarian adult craftwork. Children's flower lore has a wit and inventiveness that is worth cherishing in its own right. And plant use is so charged with make-believe that it would be foolish to try and draw a hard and fast line between what is play and what is practicality. (In which camp, for instance, would you place a corkscrew walking stick or a bibleleaf?) So since games with plants do have an element of pioneering and exploration about them, I felt it was excusable to include a section on them here. Which is already giving the subject far too serious a tone. Plants are fun, and that is a good enough use for anything.

A lot of their richness would be lost by simply cataloguing the games, particularly the glimpses they give of children's inner lives, and their experience of place and season. So I am very grateful to Kathleen Hersom for permission to reproduce extracts from a marvellous first-hand account she wrote of the plant games of her North-Country childhood early this century. ('Games with Flora', *The Countryman*, Autumn 1973). Much of what she describes catches fascinating echoes of the uses of plants described elsewhere in this book. She begins with high summer, in the harvest fields:

The dollies that I remember from the cornfields were very much simpler than the sophisticated corn dollies of the modern revival. They were the little ballerina poppy dollies that our mother taught us to make from the scarlet poppies that brightened so many harvest fields; much admired by us, but hated by the farmers.

Tiny features were marked with a pin or a pencil point on the green seed capsule which wore its own natural flat ribbed cap; a few stamens were removed in front of the face, the remaining ones forming an impressive Elizabethan-type ruff. The petals were turned down and tied tightly at the waist with cotton or a blade of grass. A length of stalk was pushed between the petals for arms, and, if fussy, another piece could be pushed up into the waist to make a second leg – but as this leg was inclined to fall out with use, my dollies were usually born one-legged, and one-legged they remained. Their lives were butterfly short, but full of incident.

In the cornfields too, we found the scarlet pimpernel – our toy barometer – wide open for sunny days, and shut tight for rain; it was at least more reliable than the time-keeping of our dandelion clocks, or the buttercup test for liking butter. How credulously the very young trust to cherry stones and daisy petals to show the future, or reveal the unknown; 'Tinker, tailor, soldier, sailor', and 'He loves me, he loves me not' are the chants.

The poppy dollies were followed, in autumn, by the acorn men, and various unidentifiable birds and beasts made from fir cones and horse-chestnuts. Carefully following the instructions given in the current copy of the *Rainbow*, we chose a small acorn with a cup to make a head and cap; then turned the cup uppermost, and fixed this with a pin to a larger body acorn. Limbs were made with pins, or match-sticks. They were not a great success; either the acorns were too hard, or our fingers not strong enough to fix the pins firmly. But I do recollect improving on the model by laboriously sticking one of the bearded sepals of the sweet briar to my acorn man's chin.

There is no limit to children's ingenuity with the bits and pieces of the plant world. Teasel heads make miniature hedgehogs, and foxglove flowers turn into finger puppets. Children in the Norfolk fenlands used to roll back the white felt of coltsfoot leaves to make shiny-green, fur-edged mirrors. In North Carolina they make turkeys out of pine cones. The cones are roughly shaped with a penknife whilst they are still moist and shut, and then arranged near a fire.

Some plant toys: a plantain gun, spring kittens (made by gluing pussy willow and hazel catkins onto paper), poppy doll, acorn man, conker cat and rose-hip necklace.

As they dry, the scales open and ruffle up like a turkey's feathers.

Then there are the multitudes of seeds and fruits, whose natural means of protection and distribution provide yet more ideas:

We were told of a wicked game in which a certain uncle removed the hairy itchy seeds from rose hips and dropped them stealthily down the back of a companion, to the intense discomfort of the victim and equally intense enjoyment of sadistic spectators. We were told we must never play the game, and surprisingly, I do not think we ever did.

Another unpleasant game was that of removing the husks from a head of timothy-grass and twisting the naked stem backwards and forwards in little girls' long hair: combing the knots out afterwards was agony. Burrs from the burdock were even worse. Seeing how many 'sticky-bobs' from goosegrass or burdock could be stuck on someone's back before they retaliated was legitimate, although even this mild sport was frowned on by a few tweedy grown-ups who claimed that it ruined their clothes. There were restrictions too on pea-shooters made from the hollow stems of what I think must have been hogweed.

We went visiting decked with bracelets made from twisted rushes, and bright with beads of rowan berries, haws and rose hips; and, long before the hippy fashion, we had our necklaces of melon seeds. We puffed our acorn cup pipes, and in high summer

we paraded up and down beneath the enormous umbrella leaves
of the butterbur that grew along the riverside. . . .

As well as the traditional games of 'conkers' there were other
competitive games that we learned or invented; one was a behead-
ing game where we each had a head of ribwort plantain with which
we took turns to knock our opponent's head, the winner being
the one whose plantain endured the longest. We leaned over the
bannisters dropping winged sycamore seeds down the staircase
well to see whose 'aeroplane' reached the bottom first. We split
hard rushes *Juncus inflexus* along their entire length, and then,
with our thumb nail, pushed out a squirming white worm of pith,
which we measured to see who had the longest candle wick. And,
of course, for centuries before it was christened by A. A. Milne,
there must have been games of 'Pooh sticks', varied only by the
materials used. . . .

In story books we came across children (mostly shepherd-boys)
who made wonderful pipes from hollow stems of elderwood, on
which they played beautiful tunes – but in reality I never came
across any such little musicians. The best I could do was to blow
on a bazooka made from a folded privet leaf which was held by

There are plants which can be used as 'clocks' in a more literal
sense than dandelion. Goatsbeard (right), often known as
jack-go-to-bed-at-noon, opens early in the morning and
shuts at midday; scarlet pimpernel (left), shuts at about 3
p.m. on clear days. Linnaeus constructed a kind of floral
sundial from such light-sensitive flowers – 'a clock by which
one could tell the time, even in cloudy weather, as accurately
as a watch'.

first fingers and thumbs and waggled to make a splendid sound. I blew on blades of grass as well, but was discouraged by cutting my lip so often. With practice a good loud report could be made by pinching shut the bladder of a bladder campion and bursting it smartly.

The only sadness in all this is the decline in the numbers of so many of our wild flowers, which means that we must encourage children to be frugal in their picking. My own feeling is that gathering flowers for games helps develop a child's sensitivity towards plants and their needs. But with already scarce species like the cowslip it is an indulgence we can no longer afford, though let's hope the situation never becomes so bad that we have to decide, as Kathleen Hersom puts it, 'how many inches of daisy chain we can allow per childhood'.

Elder

In the section on hedges in Thomas Hale's *Compleat Body of Husbandry*, 1756, the elder is spoken of in the kind of terms usually reserved for exotic spice and timber trees:

The elder is the quickest of any in its shooting; and it will bear planting so large, and takes root so easily, that it may be called an immediate fence. To this let us add, that the flowers and berries bear a price at market; and that the wood of the old stumps is valuable, and of sure sale to turners: and we shall find that there is great reason for naming the elder among the hedge shrubs, for that it equals any of them in value.

Two hundred years later a book on traditional country crafts has just a dozen words to spare on the shrub's place in hedgerow economy: 'The unwanted wood such as elder and bramble is cut out.' It is a sad decline in reputation for one of our most useful shrubs, but elder's image has always swung between these poles of veneration and distaste. Even its anatomy is ambivalent. It is too large for a bush yet too small to be a tree. Its wood is as tough as ebony, yet the young branches are composed almost entirely of insubstantial pith. The flowers smell sweet, the leaves like old mice nests. It is a companion of death and decay, growing as fast as a weed near dung heaps and graveyards, and killing off most plants under its shade. Yet its fruits make one of the most soothing of all home medicines – not a life-saver perhaps, but at least a healer.

A witches' plant, or a sacred one? Perhaps it was an inquisitive desire to resolve these contradictions that led to the elder being so thoroughly explored, and having had more practical uses discovered for it than any other temperate zone plant. There is not one single part of the elder, from root to bud, that does not have an authentic economic use. I have discussed many of these individually in other parts of the book; but I felt it was worthwhile bringing them together here to show the extraordinary versatility which can be uncovered in just one plant.

There are several species of shrubby elder, all similar in form and structure. The common European and American elders (*Sambucus nigra* and *S. canadensis*) are very similar shrubs, both growing to about 5 metres in height, with furrowed, corky-barked old wood and pith-filled new, leaves formed usually of five, slightly-toothed opposite leaflets, flat umbels of creamy-white flowers in June and clusters of purple berries in early autumn. Those are the shrub's basic components. They also share a name; and even at this point of beginning – the plant's christening if you like – we find the blurring of magic, religion and secular common-sense that marks the whole of elder's social history. The Old English root of the word is *ellaern*, and the Anglo-Saxon, *eldrun*. These most probably derive from the Anglo-Saxon *aeld*, meaning fire, because the hollow branches were used like bellows for blowing on fires. The word becomes compounded in the name of the Scandinavian spirit Hylde-Moer, the guardian of the tree whose permission you had to seek if you wished to cut the wood.

Whatever role the hollow branches played in forming elder's names, cleared of their pith they must have been one of the shrub's earliest useful products. They made flutes and pipes as well as a primitive bellows tube. Pliny mentions the belief held by some peasant people that the most sonorous pipes were made from elders which grew out of earshot of a cock's crow.

Even in an unworked state, elderwood was regarded as a magical timber. It carried the dubious but undeniably powerful spirit of the plant. Drovers would use a switch of elder to protect their cattle from disease and injury. Up until quite recently hearse drivers are said to have favoured elder handles for their horse-whips. It would be difficult to conceive of a more perfectly condensed symbol of centuries of jostling between superstition and practicality: a handle shaped out of a hard and durable wood, that was an amulet for the driver with his dangerous cargo, and, with the supposed insect-repellant qualities of the plant, a kind of fly switch for his horses.

With its combination of stinking leaves, collapsible branches, but indisputable virtues, elder came to be looked on as a kind of vegetable lamb dressed up in shabby wolf's clothing. It cried out for a myth to account for all these paradoxes, and it was not long before one was supplied. In the Middle Ages it began to be suggested that elder was one of the woods of the Cross – which sanctified or cursed it, according to the writer.

> Bour-tree bour-tree: crooked rong
> Never straight and never strong;
> Ever bush and never tree
> Since our Lord was nailed on thee.

went one old Scottish ballad (bour means 'pipe'). To complete the symmetry of the story, Judas chose an elder tree to hang himself from, balancing up whatever blame or blessing had fallen on the shrub as a result of its earlier role. The latter part of this legend gave rise to one of the most far-fetched applications of the Doctrine of Signatures. The fungus *Auricularia auricula*, which grows almost exclusively on elderwood, came to be known in medieval times as Judas's or Jew's Ear. It has the shape of an ear, but a surface more like a mucous membrane; and by a complex of associations – the stretched neck of the man hanged on the host tree, the resemblance of the fungus to a tonsil, perhaps – came to be a specific, *'fungus sambuca'*, for sore throats.

But most of elder's uses in early medicine were less outlandish than this. The inner bark was brewed up as a purgative. The green leaves made an ointment for bruises and chilblains. The umbels of creamy flowers (which John Gerard delightfully described as carried on 'spokie rundles') were taken against lung disorders and infectious skin diseases. The wines and robs made from the berries were one of the oldest and most popular country brews for coughs and colds.

So varied and extensive were the uses of the elder that John Evelyn wrote in 1664 that:

If the medicinal properties of the leaves, bark, berries, etc, were thoroughly known, I cannot tell what our countryman could aile, for which he might not fetch a remedy from every hedge, either for sickness or wound.

Sylva, or a Discourse of Forest-Trees

For the berries he reserved a piece of typically extravagant euphony:

an extract . . . is not only efficacious to eradicate this epidemical inconvenience [scurvy] . . . but is a kind of catholicon against all infirmities whatever.

Evelyn relied for much of his information about the elder on a book which had been published some twenty years before, *Anatomia Sambuci: or, the Anatomie of the Elder*, by the German physician Martin Blochwich. In 230 pages of closely detailed eulogy Blochwich prescribes parts of the elder for more than seventy distinct diseases, from toothache to the plague. Here is his recipe for a 'mucilaginous anodyne liquor':

Of quick snails, newly taken out of their shelly cottages; of Elderberries dried in the oven; and pulverised; and of common salt, of each as much as you will put it in the straining bagg, called *Hippocrates* sleeve, making one row upon another, so oft as you please; so that the first be of snails, the next of salt, and the last of berries, continuing so till the bagg be full; hang it up in a Cellar, and gather diligently the glutinous liquor that distils out of it little by little.

He also quoted the old belief that a tincture made out of the blossoms could restore sight to the blind. A considerable claim; yet elder water, as *Aqua sambuci simplex*, had a place in the Pharmacopeia as a soothing eye-lotion until the late 1950s. It is still one of the finest skin cleansers (a refined version is sold as *Eau de Sareau*) and very easy to make. Simply steep a cupful of the petals (preferably shaken from the stalks and dried) in a litre of hot water for a couple of hours, adding a little honey, witch-hazel or glycerine, as you like. It has the sweet smell of muscat grapes, and will keep for longer than many plant lotions.

In fact many of these old elder recipes, inspired though they were by magic, have been translated into modern counterparts. The hollowed-out stems make pop guns. The pith inside them is one of the softest, lightest solids known (its specific gravity is $0 \cdot 09$, against cork's $0 \cdot 24$). It is used commercially for gripping small biological specimens whilst they are cut into sections for microscopy. It is also absorbent enough to be used by the makers of watches and scientific instruments for dabbing oil and dirt from delicate mechanism. By contrast, the wood of the mature trunk is so hard and pale that it makes a fair substitute for box. Elderwood is still used occasionally for making cogs in mill machinery, pegs, butcher's skewers and combs, and whilst still green is one of the best woods for carving (p. 30). Even badgers have a cosmetic use for the wood, sharpening their claws on the corky bark.

A sprig of the leaves, worn in your hat, will keep the flies at bay. An infusion of the leaves will even kill some of them off. (Some writers have recommended wearing this lotion in preference to the sprigs; but I fear this would be likely to repel your friends along with the insects.) The leaves also

make a green dye when mordanted with chrome. The dye
from the berries, on the other hand, is a fast blue-black, and
has been used for colouring hair as well as fabric.

The culinary uses of the fruit and flowers are perhaps the
best known of all. The blossoms can be used for flavouring
milk, stewed fruit, cold drinks and gooseberry jam; or can
be dipped in batter and fried. The berries can be turned into
wine, chutney, ketchup, and jelly.

It is a formidable list of uses, and most probably an
incomplete one, when you consider how comparatively
neglected elder is now as an economic plant. And consider
for a moment the complementary list, elder's *requirements*
for producing this abundance of produce. It will tolerate the
most extreme pollution and the worst sorts of weather. You
will find it on railway embankments in industrial estates,
and on clifftops lashed by salt gales. Being fond of
nitrogen-rich refuse, it is also happy to live conveniently
close to human dwellings (it has a particular taste for
drains). It grows extraordinarily fast, sometimes putting as
much as two metres on its branches in a single season. The
flowers are largely self-pollinating, and the seed-bearing
berries eagerly gobbled and dispersed by birds. As G. H.
Knight once pointed out, there could scarcely be a living
material more appropriate to our age, and to an environment in a constant state of upheaval and change. It is
cheaply and rapidly produced, and either quickly repaired
when damaged or scrapped and replaced. And in full
bloom, against the grey stone of a hill or a church wall, it has
a pale and ethereal charm all of its own.

When the accounts are drawn up like this, elder emerges
not just as economic but as truly economical, a plant which
makes the best and most efficient use of its resources.
Mysterious, fecund, marvellously adaptable – nothing
about it is dull, and none of it need go to waste.

Faggots

A faggot is simply a bundle of brushwood, used normally to start a fire. But the term is also applied to any collection of poles of roughly similar length and thickness which, for the sake of convenience, have been bound up into a twiggy sheaf. It is a versatile contraption, as useful for building up paths over muddy ground and providing nest sites for wrens as for building fires.

You can make your own faggots by gathering up hedge trimmings and sorting them into different thicknesses as you go. These are an incidental waste product these days, though it is not so long ago that hedges were deliberately cut for the twigs they provided. But the basic traditional source of useful brushwood was the coppice, a class of managed woodland that has sadly almost vanished from the countryside.

Our current attitude towards the production of usable wood is much the same as it is towards arable crops. We plant a tree – for one purpose and one alone – let it grow, cut it down, dig it up, and plant another. It is an extraordinarily thriftless and inflexible practice, wasteful of time and space, and exhausting to the soil. It is an ironic comment on the short-sightedness and lack of imagination of our economists that we need to go back some hundreds of years to find a home-grown working example of the supposedly new ideas of renewable resources and continuous yields.

The old coppice wood, which survived as the principle source of wood products for well over a millennium, was a model of economic and ecological common sense. Coppices were encouraged to develop a variety of tree species, for the simple reason that the community needed a variety of different woods. The best trees were allowed to grow into large specimens for timber, and the remainder cut back to ground level every seven to twenty years on a rotation system. Cropped like this, the bases, or 'stools', sent up progressively more young shoots every year, and the life of

A recently-cut coppice 'stool'.

the tree was actually prolonged, being relieved of the stresses of overweight and decay. There are still some living – and productive – ash stools in eastern England which are over three metres wide, and probably date back to medieval times.

The uses found for this continuous supply of mixed underwood were enormous. The thinner twigs were used for firewood, thatch, and as pea-sticks. Wands from hazel were woven into hurdles, and oak poles cleft for basket work. Chestnut coppice made fencing, and the thicker pieces of field maple were turned into kitchen utensils. Ash, one of the most productive of all coppice woods, was used for a multitude of farm implements (see p. 40). Nor were timber and brushwood the only products of the coppice-with-standards woodland. Nuts were picked from the hazel bushes in the years when they were not cut (up to 30 cwt an acre). Pigs were sometimes allowed into the woods to feed off acorns from the timber oaks. Almost nothing was

These nut-crackers, fashioned from a single piece of hazel-wood, were made by a Sussex hurdle-maker in the 1930's. He used to carry a pair when working in the hazel coppice in the autumn. After shaping a piece of straight wood with his knife, he soaked it, doubled it over and then bound it tightly with a strip of split hazel until it dried out. Although in some ways these nut-crackers are an exceptionally ingenious piece of work, they are also typical of the on-the-spot improvisation of the old woodcraftsmen.

allowed to go to waste. The accounts books of the old woods are full of references to monies received for the by-products of the coppicing itself: woodshavings, trimmings, bark and leaves (for dyes), even rotten wood – all of them grouped together under the graphic dog-Latin phrase 'loppium et chippium'. And no doubt even the most prosaic villagers would not to be averse to cutting a few pussy willow fronds for their cottages in early spring.

Luckily a few old coppice woods are being saved as nature reserves and working museums. But why not keep the practice alive in gardens as well? It is a far preferable way of dealing with a hardwood tree that has outgrown its suburban garden than tearing it up by the roots. Young willows and poplars respond particularly well. Simply cut them back in winter to about 10 cm above ground level, and by summer they will have put up a sheaf of new shoots up to two metres tall. You can cut a proportion of these whenever you wish; in winter for carving into pegs or fishing floats, or as palm at Easter.

Firewood

Wood – live, dead, roasted to charcoal or fossilised to coal – has always been man's most important domestic fuel. Is there anyone who, given the chance, would not choose a log fire to heat and brighten a room through the winter? No other fuel draws people together so well on a cold night. The type of wood is not important; all that matters is that it burns.

But different trees do burn differently: fast, smouldering, bright, hot, smokey. Some are best for kindling, others for keeping a fire in overnight. If wood was your chief or only source of fuel (as it was for most country dwellers) you needed to know which was which. So all manner of woodcombers' saws developed – proverbs, rhymes, legends, mnemonics and epigrams, whose purpose was to act as memory jogs and shopping lists for those who had the daily chore of gathering the wood for stove and hearth. The first I came across (as a young Boy Scout) was so adamant in its do's and don'ts that I assumed that they were regulated by pagan taboos as much as the laws of combustion. Here is another which is more practical and less evangelical:

> Logs to burn! logs to burn!
> Logs to save the coal a turn!
> Here's a word to make you wise
> When you hear the woodman's cries.
>
> Beechwood fires burn bright and clear,
> Hornbeam blazes too,
> If the logs are kept a year
> To season through and through.

Oak logs will warm you well
If they're old and dry,
Larch logs of pinewood smell
But the sparks will fly.

Pine is good, and so is yew
For warmth through wintry days,
But poplar and willow too
Take long to dry or blaze.

Birch logs will burn too fast,
Alder scarce at all.
Chestnut logs are good to last
If cut in the fall.

Holly logs will burn like wax –
You should burn them green.
Elm logs like smouldering flax,
No flame is seen.

Pear logs and apple logs
They will scent your room,
Cherry logs across the dogs
Smell like flowers in bloom.

But ash logs, all smooth and grey,
Burn them green or old,
Buy up all that come your way
They're worth their weight in gold.

In this case poetry is borne out by science, and the advice above can scarcely be improved on for those who would like 'to save the coal a turn' in their own hearths. Birch, pine, and increasingly elm, are the timbers you are most likely to be offered for sale as firewood, and are amongst the commonest to find in woodland debris. Yew and pear will be the rarest, unless you are lopping trees in your own garden, in which case you might also have some lilac, one of the most aromatic woods there is. You could become a specialist in scented woods, like John Wyatt, who lived off the wild in a hut in England's Lake District:

I soon became a connoisseur of wood smoke. For fragrance, in my opinion, there is little to match juniper; I would stack the wood aside against the days I had visitors. Apple and well-seasoned cherry are pure luxury too. Holly and birch have a clean tang.

Ash, particularly green ash, smells like washing day. Old oak has an honest, pungent lusty smell as you would expect. The other hardwoods are hardly worth a mention smokewise. The soft woods; pine, spruce or larch are rather vulgar; but there is something to be said for a really old vintage larch root.

Once one gets the taste for smoking wood it is possible to mix and obtain subtle flavours; and invent recipes. Prepare a fire base of larch kindling; add well-seasoned oak until the logs redden deeply; place one large back-log of holly, and add, from the fire back to the front, one crab-apple log, one of well-dried cherry and one of birch. An ideal after-dinner mixture.

The Shining Levels, Geoffrey Bles, 1973

Similar firewood cocktails are sometimes used in smoking fish and meat. Juniper, particularly – too scarce and spindly to use on its own – is often added to give a winey tang to smoked salmon, though the most exclusive fish is cured with cedarwood. Other coniferous trees aren't often used, as they deposit too much resinous material on the fish. Normally it is just a single species of hardwood, selected because of the affinity between its scented smoke and the food: oak for ham, birch for haddock, apple for oysters.

Most of us cannot be so fussy about the woods we burn, and gather up what we can find. There are a few rules worth following here. Don't cut live firewood from other people's trees (or from your own, except in conditions of dire need; there are much better uses for it). Dry dead wood is the best. In most countries there are ancient rights to the gathering of this kind of firewood. In England, for instance, a country workman had the right to take as much dead wood as he could from a tree 'by hook or by crook' (the origin of that phrase of determined ingenuity), that is by shepherd's crook or labourer's weeding hook. It would need a patient and scholarly lawyer to determine whether this right of 'Mesyryd' still had force today; but few landowners object to the gathering of dead wood from the ground, and a small amount from the trees (which is actually a service to the growing branches).

And don't forget other sources of waste wood: abandoned hedge and garden prunings, and the shavings that build up when you are shaping wood. A sense of thriftiness is especially important when your raw material is literally going up in smoke. I still get angry when I remember the squandering, joyless bonfires that followed the great gales of January 2nd, 1976. It would not surprise me if there were more trees blown down across north-west Europe that night than are deliberately felled in a whole year. In some areas whole plantations were razed to the ground. But if it was a disaster, it was a natural one – the rather precipitate return of these trees to the normal cycle of life and death inside a wood. Which is more than one can say about the fate of those that were unfortunate enough to fall on public land. It was clear that local authorities had to act, and act quickly. Some villages were completely cut off, their telephone links severed and exit roads blocked. But what kind of fussy, municipal miserliness made them burn up so many of the fallen trees on the spot, in great blazes that lingered for days on the edges of empty recreation grounds and deserted trunk roads, with not a solitary impromptu barbecue to take the edge off the futility? Would it really have been so much more expensive and time consuming to have stacked the logs up as a free community woodpile, a small gift of winter *firebrote* from our modern feudal lords? The irony and waste were underlined a fortnight later, when a long spell of fierce cold began, and the soaring cost of coal left millions of old people shivering in their homes.

Indoors or out, use the thinnest, dryest twigs for starting your fire. Hawthorn, larch and birch are the best, arranged in a criss-cross pattern over your paper or kindling (see the entry on tinder, p. 150). For a fast blazing fire, stack thicker branches as nearly vertical as you can, propping them up against each other like the frame of a wigwam. This takes advantage of the tiny air and sap channels that run *along* the wood and act as internal chimneys. For a slower, hotter fire, lay a few much thicker logs *across* the grate, with their grain at right angles to the flames.

Finally, if you should ever need to beat out a woodland fire, remember that most foresters still, paradoxically, use the same material it is best to start one with – a bundle of fine birch twigs.

Flaxes and fibres

Fibres occur in all but the simplest of plants. They are specialised cells that give strength and pliancy to tissues, and are the reason that stalks are able to bend without snapping. Their value to humans is that they can be spun together into threads, and it is consequently the plants which possess the longest and most elastic fibres that have been sought after. Wood fibres, for instance, are no good for spinning because they are too short (though they can be *pressed* together for paper). Straw fibres are too brittle. But cotton fibres (from the downy seeds) are single cells three thousand times longer than they are thick.

Some of the finest fibres for spinning come from plants in the flax family. Their qualities were realised very early. Traces of flax have been found in prehistoric Lake Villages, and a kind of linen was used by the Egyptians for shrouding mummies as early as 4000 BC. Their flax was the common species, *Linum usitatissimum*, which has been cultivated and naturalised for so long over such an area of the globe that its original home has never been discovered, though it was probably somewhere in central Asia. The elaborate preparation needed to convert the plant into usable fibre (not unlike that for woad, see page 21) is also a testament to the sophisticated understanding of plant anatomy which had evolved even by prehistoric times. The stalks are picked in late summer, dried, stacked, and then thoroughly wetted, so that over the next few weeks they start to rot. The stems are then dried again, and *scutched*, that is beaten to remove the rotten tissues from the tough skeletal fibres. In some peasant areas this is still done by hand – or rather by arm, the bundles of stems being flexed and whipped like carpet-beaters. The resulting fibres are pale yellow, between 50 and 100 cm long, and very strong, particularly when wet. Pliny, in his *Natural History*, was quite breathless about their virtues:

What department is there to be found of active life in which flax is not employed? And in what production of the Earth are there greater marvels to us than in this? To think that here is a plant which brings Egypt to close proximity to Italy! . . . . What audacity in man! . . . . Thus to sow a thing in the ground for the purpose of catching the winds and tempests; it being not enough for him, forsooth, to be borne upon the waves alone!

The flax which the Romans grew was almost certainly pale flax, *Linum bienne* (a species with more lilac-tinged blossoms than *L. usitatissimum)* which is native in Britain and the Mediteranean. Its cultivation continued after the Romans had left, and so did the power of flax to evoke literary effusiveness. In the mid 13th century, Bartholomaeus Anglicus, one of the first English writers on plants, includes a marvellous passage on the preparation of linen in his *De Proprietatibus Rerum*, describing the soaking and drying of the flax, its binding in 'praty bundels', and how it was subsequently 'knockyd, beten and brayd and carflyd, rodded and gnodded; ribbyd and hekyld and at last sponne', and how the resultant fibre – surely as fine as silk after such a variety of exotic treatments – was made into fish nets, sails, ropes, sacks, sheets and shirts. He concludes: 'none herbe is so nedefull to so many dyurrse uses to mankynde as is the flexe.'

The uses of flaxen thread have scarcely changed at all in the succeeding seven centuries. But I have a feeling that the eulogies of these writers may have had something to do with the heartwarming sight of a field shot through with flax's sapphire blossoms, as well as with the enduring usefulness of its fibres. One of England's finest modern country writers, John Moore, based a whole book (*The Blue Field*,

Collins, 1948) on the brilliant challenge its cultivation symbolised to drab Ministry-sponsored monocultures.

In the wake of the oil crisis, and the rising cost of synthetic fibres, we may begin to see flax fields about our countryside again. Meanwhile, neither naturalised nor native species are common enough to make the production of wild linen a viable proposition. But it is possible– though difficult– to make acceptable yarns from nettle fibres, using a similar process (though nettle's fibres are much shorter). Nettle cloth is still manufactured in northern Europe, and the plants actually cultivated for the purpose in some areas. When Germany ran short of cotton during the First World War, they used enormous quantities of nettle to make military clothing. Something like 2½ million kilograms of plants were gathered from the wild – though it took 40 kgs of these to make one single shirt!

Fleabanes

Banes, in the plant world, are not just pests and annoyances. They are outright poisons, especially to valuable human familiars. Bitter experience no doubt named the henbane, the dogbane (a middle eastern *Periploca* vine), the horsebane and sowbane; and rather more deliberate slaughter, the wolfsbane (our garden monkshood). But the animal, as they say, is irrelevant. Most of these plants are toxic to anything that eats them. A single bite of the root of cowbane (a sister plant of the infamous hemlock) can bring on fatal convulsions in man and beast alike. But here and there you will come across a banewort, or 'Devil's Bane' (St John's worts were called this in Somerset), whose powers of despatch were clearly believed to be due to more magical, deterrent qualities.

Fleabanes are neither out-and-out insecticides nor wholly products of such wishful thinking. Their original role was as fumigants, and they were burnt inside rooms to drive out fleas and lice. The common fleabane that grows in damp grassland in Europe *(Pulicaria dysenterica)* was the species most commonly used. It is a pleasant plant, with furry, wrinkled leaves and tight yellow flowers that often scarcely over-top the grass. It reminds you of an old fashioned pot marigold, and bearing in mind its favourite damp haunts, I can't help thinking that the name marsh marigold would have sat more happily on its daisy-flowers than it does on that regal buttercup, *Caltha palustris.* Still, a fleabane it is, and if you crush the leaves and catch its strange scent of cats and carbolic and chrysanthemums it is easy to understand why it was believed to repel fleas. But insects' attitudes towards scent would be most unlikely to be the same as ours (many smell through their feet!) and as we shall see later, it is the fleabane's family and biochemical ties with the marigolds which may supply the clue to its insecticidal powers. (The most famous plant insecticide, pyrethrum, also comes from a member of this family, though it is not a wild plant in the region covered by this book. Pyrethrum is the dried and powdered flower-heads of the insect plant, *Chrysanthemum cinerariifolium*, indigenous to south-eastern Europe, though now cultivated in many parts of the world. It has an almost instantaneous stupefying effect and has found an honoured place in many proprietary insecticides.)

Other fleabanes, and various species of *Inula* and *Conyza*, have also been burnt to fumigate rooms. So have mugwort and wormwood, though the pungent sprays of the latter were more often simply strewn on the floor:

> While wormwood hath seed, get a bundle or twain
>   to save against March, to make flea to refrain:
> Where chamber is sweept, and wormwood is strewn,
>   no flea, for his life, dare abide to be known.
> *Five Hundred Pointes of Good Husbandrie*
> Thomas Tusser 1573.

With the wormwood family we are into a group of more immediately useful insect repellants. Southernwood, or lad's love, *Artemisia abrotanum*, has long been hung in wardrobes to keep away the moths. In France they call it *garde robe*, plain and simple. The French are lucky enough to have it as a native in the south, but it is also naturalised around dwellings in North America and a favourite in herb gardens everywhere for its feathery grey-green foliage. Try a spray amongst your own clothes. It may not deter the moths as much as you would wish, but you will find its fresh lemon-and-lavender scent a good deal more pleasant than camphor.

Try, too, a sprig of wormwood or southernwood in your hat to keep away summer flies. Elder, the acrid, and gale, the sweet, tucked behind the ear, will also keep the midges down (though not completely away), demonstrating again that insects' tastes in perfume bear no relation to our own. A spray made of an infusion of elder leaves in water (1 kilo to 2 litres) will also kill aphids, and a similar decoction of the fly agaric toadstool has been painted on walls to drive away house flies (see p.154). To complete the list of oddities, there have been two vegetable fly papers. William Withering reports that the glutinous spring leaves of the alder were strewn over floors; and in Spain they still sometimes hang the sticky roots of elecampane around their doors and windows.

It would not be unreasonable to explain the ability of many of these plants to kill and deter insect predators in terms of some kind of chemical defence mechanism – a built-in antibiotic. And in recent years some exciting collaborative work between gardeners and botanists (a long overdue reunion: the two arts were aspects of a single profession when fleabanes were first named) has proved that such a mechanism does exist. The hero of the story is the 'African' marigold, *Tagetes erecta*, a three-metre-tall giant that belongs to the same group (Compositae) as the fleabanes and wormwoods we have already met. It is not African, but Mexican; and it is scarcely even an occasional weed in Europe. But I believe it deserves a place in this book. What has been unravelled about its power over any nearby organisms with aggressive pretensions – be they animal or plant – has implications for the whole of agriculture.

The clues which led to what has become known as the Tagetes effect were a series of seemingly unrelated mysteries. Why did representations of other *Tagetes* (useless as a food) so often accompany pictures of maize and beans on the vases of pre-Inca farming civilizations? Was there any scientific basis for the widespread belief in 'companion' plants, in the power of one species to influence the growth of others nearby? How did underground organisms affect each other at a distance? They plainly did in the case of potato eelworm larvae for instance, for the eelworms will not hatch out unless the field is planted with potatoes. If it is planted with peas or sugarbeet, the larvae stay in a resting state.

Work at a number of plant research stations showed that it was secretions from the roots of potatoes that made the eelworms hatch. No other root secretions would do this – though pea roots would make a pea eelworm hatch and move towards its eventual source of food. These secretions are a kind of underground scent, as distinctive as body odour; and the reactions of creatures to them are as specific and sensitive as those of male and female moths which can sniff each other out over a distance of hundreds of metres.

It was a Dutch nurseryman, Berg-Smit, who first uncovered the possibility that other root secretions might work against eelworms. Planting *Tagetes erecta* as a cut flower crop after removing his daffodil bulbs, he found that he drastically reduced the damage caused by the narcissus eelworm. He reported his findings to the Wageningen Research Station, who immediately began detailed investigations. They tried different species of *Tagetes* in different conditions, and found that many of the marigolds would kill eelworms up to a metre away from their roots. The research workers were even able to identify (and later synthesise) some of the complex chemicals which made up the subterranean exhalations. The maximum effect of the plant is always on those creatures that have selection mechanisms to pick out their own food plant by its root secretions; *Tagetes* seems to scramble this mechanism. But it has no effect at all on beneficial scavengers like worms that live off dead plant material.

An almost perfect insecticide, then. But there was more to come. Experiments with *Tagetes* species in gardens and laboratories all over the world showed that they could also kill the underground spores of certain fungus diseases, and reduce the growth of expansive weeds like bindweeds and thistles. Almost always the strongest effect came from the wild Mexican *T. minuta*, the plant so often figured in the sacred art of Central and South American agriculturalists. Was it deliberately encouraged on their terraced fields of potatoes and beans, and sanctified because of the protection it gave to the crops? A rotation system which included *Tagetes* (even just as a tolerated weed) could well have enabled these farmers to grow the same crops on the same land for hundreds of years without the infestations that plague modern arable monocultures.

What has all this to do with our own wild plants? One answer is that the active chemicals in the *Tagetes* root secretions (chiefly thiophenes) have also been identified in many of the more lowly fleabanes we looked at earlier. So one, at least, of the old herbalists' apparently quaint recipes could well have a sound scientific basis. But I feel that the greater revelation is the immense and complex network of chemical activity which goes on beneath the earth. The soil is where a plant does the bulk of its living, where it takes in food, fights for space and struggles against a legion of predators that dwarf in number and variety those that attack its foliage. It is perhaps no wonder, then, that over millions of years some plants should have evolved highly sophisticated techniques for self-defence.

It is hardly necessary to emphasise that, with the world shortage of food, and the damage we are inflicting on our environment (and ourselves) by our reckless use of synthetic pesticides, any lead of this kind is worth following up. The screening of all manner of plant secretions for their effect on particular insects and weeds would provide a magnificent opportunity for professional plant chemists and amateur naturalists and gardeners to work together.

Forks, in trees

The fork in a branch, or trunk, is one natural form that man cannot reproduce synthetically. We can copy its shape, of course; but not the minute gradations of cell length, the stretching and bunching of bark and fibre that produce not just the marvellous turbulence of grain around a junction 'knot', but that perfect balancing of strength and shape that exists between sibling branches. The limbs that form a fork are fulfilling a biological purpose in diverging; but, mechanically, physiologically and even aesthetically, they are still part of one structure – vegetable Siamese twins, though they may soon cease to resemble each other.

Wood which has grown into a fork – or indeed any kind of angle – is always stronger than timber cut artificially into such shapes. In the timber yard of a sawmill in Staffordshire – and I am sure it is not exceptional – they keep a special pile of bent and crooked oak logs to which boat builders and the increasing throngs of restorers of old buildings come for their crooks and 'knees' and windbraces.

And even as a pure shape, what a universally useful object a forked stick is. It will make a thumb-stick, a catapult, a prop for a clothes-line or a fishing-rod, even a dowsing rod, if you have faith in such things. You can even make forks out of forks. I once saw a four-pronged hay fork in a craft shop which had been fashioned from a single length of ivy, *Hedera helix*, Europe's most abundant woody climber. Ivy proceeds up a tree or wall like an argument in complex logic; so full of diversions and divisions ('if not route A, then B'), knots, convolutions and

Alaskan fish-gaff, bound up from forked sticks.

occasional concurrences, that it is possible to find almost any kind of angle, twist or fork you could wish for by looking at just a few plants. Its main stems also have an extremely hard heartwood, and I would rate it as one of our most undervalued timbers.

But please don't go hacking ivy from standing trees in the belief that by saving the tree from an artful and ruthless scrounger, you have earned the right to it as a material. Ivy is neither a parasite nor a strangler. It has an independent root system in the ground, and manufactures its own food through its abundant evergreen foliage. All it uses a tree for is physical support, as a kind of scaffold. You have only to consider the number of ivy vines continuing to prosper on long dead trees to see the truth of this. I have seen stocky ivy bushes, trellised originally on short-lived shrub species, standing on their own multiple legs long after their hosts have rotted away. They have been known to live over a hundred years.

The inner side of an ivy branch is covered with a mat of hair-like tendrils, and these clamp themselves to a bark or wall with small adhesive discs. Only when these suckers encounter soil or deep crevices do they develop into true, feeding roots. So the only damage ivy does to a growing tree is to compete with it for nutrients in the soil and to cut off the light from some of its leaves. The impact is usually very small. And when you consider ivy's virtues, the dramatic visual effect of its dark, dense foliage on otherwise dead wood, and the cover it provides for nesting birds, you may agree that it is not a plant which should be too thoughtlessly felled.

Hay fork made from a branch of ivy.

Yet of course it is cut down, and as much by the ordinary wear and tear of weather as by misinformed humans. Most woods will have the odd fallen tree whose ivy has already started to creep across the ground in search of a new support, and which will not suffer too much from the loss of a limb. You'll need a saw to disentangle the length you choose, and knife to cut through the suckering tendrils. Then you must shave off the soft outer wood. It is worth taking your time over this. The greenwood is as spongy as balsa, and has a warm and distant scent of autumn mushrooms. The heartwood beneath is a curious glossy cream, and dries to something like the colour and surface texture of ivory. I've seen branches used to give a bleached bone effect in flower arrangements. Personally I prefer the hayfork; though it should be said that, for all its hardness, ivy wood is very porous and not too durable.

Glassworts

Glassworts are a group of seashore plants, mostly *Salicornia* and *Salsola* species, which contain a good deal of concentrated salt inside their stems. As far back as Biblical times they were burnt for their sodium-rich ashes, which were used in the manufacture of glass.

Salicornia –often known as marsh samphire– is a distinctive plant, more like a miniature prairie cactus than a native of wet and windy places. There are many different species (and some in the Mediterranean region and in North America grow by inland alkaline lakes), some stiff and unbranched, others bushy and straggling. But they are all notable for their plump, cylindrical, jointed stems, which are crisp and salty to the taste.

Glasswort is usually the first flowering plant to colonise bare tidal mudflats, and its succulence contributes towards its survival in such an unfriendly and precarious environment. The main threat to shoreline plants (as to most living organisms exposed to the sea) is thirst, the loss of their vital internal water supplies. Lashed by salt winds, or, like some glassworts, actually immersed in seawater twice a day, they are in constant danger of being dehydrated– dried out, or literally pickled in brine. Coastal plants have many defences against dehydration, and it may be that the salts in glassworts are taken in to help build up a concentration in the roots that balances the osmotic pressure in the salt-water outside.

So seashore plants act as convenient condensed packages of mineral salts (particularly sodium), and it was to win these that the early glass makers used to gather and burn glassworts, and the closely related saltworts (*Salsola* species). Formerly, large quantities of ash were imported to Britain from southern Europe and north Africa under the name of Barilla. The chief sources were *Salsola kali* and *S. soda*, and the Spanish *S. sativa*. But none of these saltworts is prolific on northern shores and when home production started in the 16th century, the more abundant *Salicornias* were used as an alternative.

After they had been gathered the plants were dried, and then burnt in large heaps. The crude ash was fused with sand to make a rough glass, of leached with limewater to make a solution of caustic soda. On evaporation this would yield crystals of fairly pure sodium hydroxide, which would make a clearer glass, or be used with animal fats for making soap.

Nowadays caustic soda, which is used in prodigious quantities in many industrial processes, is obtained from brine by direct chemical means. Yet it was those early gatherers and burners of shoreline plants who pioneered the exploration of its properties and potential.

Gums and Glues

Gums, the dictionary tells us, are 'viscid secretions of some trees and shrubs that harden in drying . . . used to stick paper together, etc.' Their function in plants is probably much the same as it is in our glue-pots: they are menders and healers. We have seen already how the resins that seep from injuries in conifer trees are a kind of antiseptic dressing for the wound (p. 47). So, probably, are the gummy exudations (edible, incidentally) of cherry trees which have been damaged by insects, and the latexes of rubber and opium plants. And if you have ever broken a dandelion stem or root you will have seen the milky juice that oozes from the edges of the cut. This, too, congeals after a short while and forms a protective scab, though in this case the scab is not brittle like dried resin (or blood – another wound-sealing fluid!) but elastic. This dandelion milk is a latex and has been used as a crude source of rubber. You can see this for yourself by squeezing some of the juice onto your hand, allowing it to thicken and then rolling it up into a small rubbery ball.

When Russia's supplies of true rubber were cut off during the Second World War, their plant scientists began exploring the possibilities of producing dandelion rubber commercially. They had a good ground base to work from. In 1931 an Asian dandelion *Taraxacum kok-saghyz,* had been discovered in the Tien Shan mountains which had up to ten percent of latex in its roots. And what roots they were, broad and pronged and penetrating as deep as 2 metres into the ground. During the war the plant was taken into cultivation in the Ukraine, and to my knowledge is still grown there. It is understandably difficult to obtain figures about the amount of land involved, but it is estimated that at the

PLATE 5.   *Medicinal plants*

1. Fennel, *Foeniculum vulgare:* waysides, hedgebanks. Parts used: seeds, August – November.

2. Water mint, *Mentha aquatica:* wet places. Leaves.

3. Dog-rose, *Rosa canina:* hedges, scrub, woods, etc. The fruit – rose-hips, September – December.

4. White horehound, *Marrubium vulgare:* dry, bare places. Leaves.

5. Elder, *Sambucus nigra:* hedges, waste places, woods. Flowers (May – July) and fruit (August – October).

6. Chamomile, *Chamaemelum nobile:* grassy and heathy places. Flowers, June – September.

7. Sphagnum moss, *Sphagnum cymbifolium:* bogs. Whole plant.

2

3

6

4

5

7

height of the war, several million acres of wild steppes were under rubber plantations. I wonder if, with good peasant frugality, they harvested the leaves too, for salads?

But for an effective wild plant glue we must go, sadly, to the bluebell (or wild hyacinth, *Endymion nonscriptus*). I say sadly because it would be wrong to uproot such a beautiful and harassed landscape plant for no other purpose than sticking paper together. It would also be illegal, since the tacky juices are in the bulb, and to procure them you would need to dig up the plants, a practice now quite properly outlawed in Britain (p.17). But if you have bluebells in your own garden and think you can spare one to share in an ancient, though probably never very important practice, dig up one of the baby-onion bulbs and scrape at it with a

PLATE 6. *Dye plants*

As a rough guide to quantities, boil together equal volumes of plant material and water – but see page 21 for more detailed notes on technique.

1. Walnut, *Juglans regia:* hedgerows, parks, plantations. Part used: the green husks encasing the nuts, July – October. Gives a dark brown colour.

2. Oak, *Quercus robur:* Woods, hedgerows. Acorns and bark give a brown to black.

3. Alder, *Alnus glutinosa:* by fresh water. The young shoots give a yellow-brown, the male catkins (February – March), a light green.

4. Bilberry, *Vaccinium myrtillus:* Acidic heaths, moors, pine woods. The berries (July – September) give a pink to purple.

5. Weld, *Reseda luteola:* dry, disturbed ground. The whole plant gives a yellow.

6. Heather, *Calluna vulgaris:* heaths, moors, mountains. The flowering tops (July – September) give a golden yellow.

7. Golden-rod, *Solidago canadensis:* waste ground. The blossoms (July – September) give a gold.

8. Bracken, *Pteridium aquilinum:* woods, heaths, hills. The young shoots (April – June) give a light green.

knife. You will gather a deposit of a thick, mucus-like slime. William Turner, writing in his *Herbal* of 1568, says: 'The boyes in Northumberlande scrape the roote of the herbe and glew theyr arrows and bokes wyth that slyme that they scrape of.' I have tried it for gluing paper together, and found that the paper sheared before the joint gave way. Geoffrey Grigson used bluebell glue to repair one of his notebooks (into which he had copied Turner's recommendation) and found his paper hinges just as firm thirteen years later.

The stickiness is due chiefly to the starch stored up as food in the bulbs, and this also found a use in laundering. John Gerard said it 'made the best starche unto that of Wake Robin roots.' Wake Robin roots – that is the roots of Lords and Ladies, *Arum maculatum* – may have made a better starch, but it was also a crueller one, full of an acrid juice 'most hurtfull for the hands of the laundresse that hath the handling of it, for it choppeth, blistereth, and maketh the hands rough and rugged, and withall smarting'. Gerard's concern fell on unsympathetic ears, and it was to be another two centuries before Starchwort was outlawed from the sweat laundries.

Heather Beds

Heather, or ling, *Calluna vulgaris*, is the dominant plant over much of Western Europe's upland moors and sandy wastes. In full flower – lilac billows as far as the eye can see – a heather moor is a glorious sight. But it is also a deceptive one, for sweeps of ling are signs of a depleted soil and a harsh climate. It flourishes in conditions that arable crops would find intolerable, and in acid ground that has been cleared of its trees. For those who share its impoverished habitats, it has been an absolute necessity of domestic life, a stand-in for a whole range of absent plants. It has been used as the foundation for wattle-and-daub walls, to thatch roofs, sweep floors and fuel ovens. Its fibrous stems have been woven into ropes, and its roots carved into knife handles (particularly for ceremonial Scottish dirks). Its flowering tops make an orange dye, a beer and a pleasantly sweet tea. Sheep and grouse thrive on the shoots, too (which is just as well, as heather moors are often the result of drastic forest clearances carried out for the benefit of these animals). And it will be no surprise to any walker who has rested in the heather that it was also universally used as bedding, for man and beast alike. It is so soft, supporting and fragrant that Scottish settlers took it to America with them, and naturalised it in a continent thousands of miles outside its natural range. How easy it is to picture those homesick Scots, taking comfort at night with the heath from their native hills.

Except in extreme climatic conditions, heaths are unstable, and tend to progress towards woodland. The factor that maintains them in most areas is grazing, which heather can tolerate but tree seedlings cannot. On moors where grouse or deer are hunted, the heather is also periodically burnt to provide a flush of especially nourishing new shoots. This, too, works against the survival of any emergent trees (though heather's root system is impervious to all but the deepest fires).

Without this kind of management most large areas of heather would vanish; and the certainty of a constant, husbanded supply was one reason why heather was such a popular material for beds. But there are also good physical reasons, which are connected with the heather's obvious hardiness. I mentioned above that heather develops most characteristically on thin, acid soils that have been cleared of their woodland cover. When the trees go, so does most of the ground's protection from the elements. The soil is

leached of its nutrients by rain, and the plants on it buffet-
ted by wind and subject to violent swings in temperature.
Few plants can thrive under such conditions. But heather
has evolved some ingenious architectural defences. Its
wood, for a start, is supple and springy, and will bend rather
than snap in a gale. Its branches tend to be short and to
cluster together in dense, mutually protective bunches.
This is chiefly to prevent the plant drying out, a constant danger in windy, shadeless places where the soil is
thin and drains quickly. Heather's tiny, slightly succulent
leaves, which roll up during dry weather to reduce their
exposed surface area, are also designed for moisture conservation.

In adapting to hostile conditions, heather has thus
evolved as an ideal bedding plant: compact, springy, and
with an almost feathery foliage. You are not likely to have
much need of a heather bed unless you are sleeping out of
doors; and in some ways the best thing to do is just to roll up
in a blanket and lie down in a thick patch. For a more formal
bed – inside a tent, say – you can simply cut short branches
and strew them on the ground. But it will be a good deal
more comfortable if you set them upright. Cut a number of
sprays about 30 cm long, plant a row firmly in the ground,
and then stack the others against these at a slight slant, until
you have built up an area big enough to lie on. What you are
building, in effect, is a mattress of heather set almost as it
grows in the ground; and its 'interior-sprung' feel is much
like that of the growing plant. So is the scent, if you are
lucky enough to pick your bed when the heather is in flower
in August and September. (Heather is so rich in nectar that
beekeepers will move their hives many miles to the nearest
large patch during the blossom season.)

Similar, and equally aromatic, beds are made in North
America from conifer branches. Young spruce twigs (softer
and springier than pine or larch) are the favourites, and are
cut, planted and stacked exactly in the manner described for
heather. A friend who has been on camping expeditions in
Canada tells me that on warm nights these spruce mattresses are often built big enough to sleep a whole family.
A convivial idea which I suspect may have been borrowed

from the Cree Indians of Quebec, who strew the floors of
their communal tepees with spruce shoots.

Both these species of springy twig make beds in their
own right. Many other materials have been used for stuffing
existing mattresses, and making a kind of palliasse, though
the result is often a good deal more comfortable than those
straw-stuffed military sleepsacks. Whilst we are in North
America, let me quote a description of an unusually oppor-
tunist mattress from the Depression era:

One evening as I entered the home of a pioneer family in the
mountains of North Carolina, not far from the Georgia line, I was
impressed by a beautiful pile of fine shavings on the floor of the
living room near the fireplace. My first thought was that they had
been used in shipping some object from a mail order house. But I
could not recall ever having seen packing material as fresh looking as these shavings were.

Upon enquiry the grandmother who was sitting by the fireplace
said, 'This is my feather bed, and picking up from the hearth a
witchhazel limb perhaps three inches through and twenty inches
long she proceeded to peel off with her pocket knife the fine curly
shavings, which when she had a sufficient quantity would be put
into a bedtick for filling.

I had been familiar with corn husks, wheat straw, and some
other materials used in country districts, but never before had I
seen anyone make a mattress of wood shavings. She assured me
that they were quite preferable to any other filling of which she
knew, having unusual resiliency and lasting a long time. It had
been a practice in their household for many years. I had the
opportunity that night to test out the good qualities of a witch-
hazel shaving mattress and discovered it excellent.
Handicrafts of the Southern Highlands, Allen H. Eaton, 1937

The number of different plant materials which have been
used for stuffing mattresses and pillows is enormous. The
most popular, for obvious reasons, have been downy seeds.
Many plants attach parachutes of silky hairs to their seeds to
help their distribution by the wind, and when these are
clustered together before dispersal, they feel very much
like down. But to stuff even a small pillow you need a
prodigious number of seeds, and it is only really practicable
to collect vegetable down from prolific seeders like willows,

thistles, cottongrasses and reedmace (*Typha* species, otherwise known as bull-rushes and cat-tails, whose splendid chocolate-brown seed spikes split open like burst pillows themselves when they are ripe). Gather the down direct from the seed-heads or catkins, if you can. It is more concentrated and accessible just before shedding, and less likely to rot than if it has been raked off the ground. John Evelyn recommends this with an early variety of sallow catkins:

The hopping-sallows open and yield their Palms before other Sallows, and when they are blown (which is about the exit of May, or sometimes June) the Palms . . . are four inches long, and full of a fine lanuginous Cotton: A poor Body might in an hours space, gather a pound or two of it, which resembling the finest Silk, might doubtless be converted to some profitable use, by an ingenious House-wife, if gather'd in calm Evenings before the Wind, Rain and Dew impair them; I am of opinion, if it were dried with care, it might be fit for Cushions, and Pillows of Chastity, for such of old was the reputation of these Trees.

Sylva, 1664

One mattress stuffing which is naturally resistant to decay is sphagnum moss – the chief component, appropriately, of those botanical curiosities known as blanket bogs. Sphagnum lives sodden with water for much of its life, and may have developed its own bactericide to stave off mouldiness. So it was understandably popular, in 'primitive' and civilised societies alike, as a basis for disposable bedding for young children.

But the oddest moss bed was reputedly made by Laplanders from hair moss, *Polytrichum commune*. This can form huge tangled mats over the ground, which can be cut and rolled up life turf. For sleeping, you simply wrapped yourself up inside a thick sheet of the moss as if it were a quilt.

Dried leaves have also found their way into beds. In England, the tape-like fronds of maritime eel-grasses, which dry to the texture of rough horse-hair, once supported a whole mattress and furniture stuffing industry. (The business collapsed in the 1930s, when the eel-grass was ravaged by disease, and related species are now imported from the Mediterranean, as 'Algerian grass'.)

Sphagnum moss, abundant on waterlogged acid soil. Every part of the plant is permeated with fine capillary tubes, making it highly porous, and twice as absorbent of moisture as cotton wool. It also contains a mild antiseptic and preservative phenols. These two factors combine to make it ideal as a wound dressing, and enormous quantities were gathered from the wild in both World Wars, and, dried and sterilised, formed a comfortable and hygienic packing for dressings. The use of sphagnum for this purpose goes back at least to the 11th century, and in the Gaelic Chronicles of 1014 there are reports that 'stricken Scots stuffed their wounds with moss'. Even deer have reputedly been observed dragging wounded limbs through the moss!

Leaves from beech and sweet chestnut have also been used. In France, mattresses stuffed with these were known as *lits de parliament* – talking beds – because of the incessant crackle they made when you lay on them.

But the bedding material which has fallen furthest from popularity is probably bracken. Bracken, or brake, is a fern with a world-wide distribution, occurring in habitats as different as the tropics and the Arctic. It is also, ironically, one of the chief weeds of heather moors, particularly where the ground has been churned up by heavy grazing. Bracken is a mixed blessing. Once established it is extremely difficult to remove, and in the green state is actually poisonous, though most animals will avoid it if there is sufficient food of other sorts. Yet once it has turned brown in the autumn it is quite harmless, and makes an invaluable and plentiful bedding for farm stock. The manure made from bracken litter is one of the richest there is, and breaks down into humus so quickly that it can be spread back on the fields without prolonged composting. So farmers scraping a living from the acid soils where bracken abounds have worked out a *modus vivendi* with the plant. They cut it with a scythe every autumn just as it is turning yellow – which keeps its spread under control; then dry and stack it as if it were hay – which provides a free source of bedding for the winter. But this annual cut-back – a cull, really, more than a harvest – is becoming an increasingly rare practice. It is now more fashionable to spray bracken with expensive weedkillers and buy-in equally expensive straw as bedding.

Yet there is still plenty of bracken about if you would like to try it as a human camp-bedding, though I should warn you that it has few of the delights of heather. It is strong-smelling when green, and uncomfortably spiky when brown. But it is still preferable, I think, to unupholstered ground.

Vegetable beddings, for men and animals alike, have considerable advantages. They are light in weight and elastic in texture. They are porous, and therefore tend to stay cool in summer and retain heat in winter. They are usually plentiful enough to be regarded as disposable if they are soiled. In the past, these qualities have made them popular for supporting other fragile objects besides the human frame. Dry bracken has been used for packing slates for lorry transport, and dried moss for padding crockery.

Hop pillows

Any of the materials I have mentioned as being suitable for rough and ready beds can, of course, be stuffed more modestly and elegantly into conventional bolsters. But pillows, being in such intimate contact with the face and nose, are worth filling with more delicate plants. Mildly scented herbs are an obvious choice, and dried fronds of rosemary, marjoram and lavender have been mixed with the usual stuffings, or slipped into the case inside a sachet. Because of the warmth and pressure of the face, herb pillows become more strongly fragrant the longer they are lain on. If their scents are mildly soporific as well as pleasant, so much the better: you may get a good night's sleep as well as an aromatic awakening. Pillows stuffed with pine needles or shavings have been popular with those whose sleep is troubled because of catarrh, the resinous vapours released from the pine providing a kind of continuous, low-temperature inhalation (p. 47). Sweet woodruff works more by suggestion, I would imagine, gradually teasing you with the impression that you are sprawled in a meadow of new-mown hay.

But the most popular ingredient for sleeping-pillows has been the hop. It's difficult to track down the origins of the hop pillow (the plant's association with other agreeable sleep-inducers may have been the start of it), but it was that renowned hypochondriac George III who popularised it. Earl Stanhope described its effect on the King's insomnia in his *Life of William Pitt*:

All questions of Regency, however, were set at rest by the King's convalescence. It is remarkable that the first favourable change was due to Mr Addington not indeed in his political capacity, but rather in his filial capacity. He remembered to have heard from his father, an eminent physician, that a pillow filled with hops would sometimes induce sleep when all other remedies had failed; the experiment being tried on the King was attended with complete success.

The 'hops' used in pillows are the same fragments of the hop plant, *Humulus lupulus*, that are used in brewing: the dried, cone-like female flowers which mature in the early autumn. If you open one of these cones you will see the source of the hop's bitter and sedative oils. Each lobe, near its base, is studded with bright yellow glands which contain a complex mixture of aromatic oils and resins known as

Hop cones, with a close-up of the lupulin-bearing oil glands.

lupulin. If you rub a patch of these glands they readily release their more volatile components. It is difficult to describe the aroma of hops in terms of any other scents. There are touches of the pungency of garlic and fruitiness of autumn apples, but with a freshness and warmth and body that is unique to the hop.

English hops are, by and large, longer than those that grow on the continent, and can grow up to 10 cm. This is because they are more frequently pollinated and allowed to set seed. The hop is dioecious, with the male and female flowers occuring on separate plants. Where male plants occur the females are often pollinated and grow bigger as a consequence. But across much of mainland Europe, where light, lager-type beers are preferred, seeded female hops are regarded as being too strongly flavoured, and the promiscuous males are hunted down and eradicated, even in the wild. As a result, continental hops tend to remain celibate, small, and mildly scented.

But the hop plant – male or female, seeded or unseeded – is nevertheless quite common and almost certainly native across much of Europe and North America. It is a climbing vine of scrubby, damp and waste places, with square stems that can reach over 6 metres in length and deeply lobed leaves. The female cones are usually mature by September, and should be picked in dry weather before they have started to turn brown. Dry them for a few days in an airing cupboard and they are ready for a hop pillow. In order to make it possible to renew the hops (they will lose much of their scent after a couple of months) it is best to sew them inside a muslin sachet about 50 cm square, rather than mixing them directly with the feathers.

There is no doubt that lupulin does have a mildly sedative action. (I have heard stories of oast-house workers dropping unconscious at their jobs, though knowing some of the ingredients of modern beers, the cause for this is open to question.) Infusions of it were included as a tranquilliser in the British and American Pharmacopoeias until comparatively recently. It is also recognised as an appetite stimulant and a mild pain-killer, and the Mohega Indians used to warm little bags of hops and use them as a poultice for toothache.

But is is debatable whether the minute quantities of the active constituents that are vaporised from a pillow have any real effect. Still, the scent is soothing in its own right, and personally I find it very pleasant to wake to a whiff of distant breweries.

Ink Caps

The names of plants rarely originate from a single source or association. Take ink caps, for instance, also variously known as inktops and inkhorns. These members of the toadstool group *Coprinus* are shaped something like inverted medieval ramshorn inkwells. But they also dissolve into an intense black liquid as they mature, and this was once used as a moderately fast ink.

The commonest species is the shaggy ink cap, *C. comatus*, which grows (often in immense colonies) in short grass, particularly on ground that has been 'made-up' with nitrogen-rich rubbish. The verge of a new road is a typical site. The caps begin to appear in late summer and continue to sprout until the first frosts. They are like small white busbies at first, but after a few days, when spore discharge begins, the pink gills darken and the flesh of the cap starts to dissolve. If you collect some caps at this stage and keep them in a bowl, you will see that they turn almost completely into liquid, which is coloured black because of the spores it contains. Although the first reference to the use of this liquid as an ink doesn't appear until 1784 (in France), it was almost certainly known long before this. Other *Coprinus* toadstools also liquefy, and it was once suggested that Indian ink that was to be used for important documents should be 'marked' with traces of ink from a particular *Coprinus* species. Since the spores are distinguishable under a microscope this would provide a way of detecting forgeries.

I don't know whether this idea was ever put into practice, but using the composition of inks as a guide to the authenticity of documents is certainly nothing new. Up until the eighteenth century most large households made their own inks. The recipes were so individual and so jealously guarded that analysis of inks has proved to be a reliable guide to the source of ancient manuscripts. Here is a recipe from the 11th century, quoted by Dorothy Hartley – '12lbs of oak galls pounded, 5lbs of gum pounded, 5lbs or less of green sulphate of iron, 12 gallons of rain water boiled each day till sufficiently done, letting it settle overnight.'

Oak-marble galls, formed by the larvae of the gall wasp *Andricus kollari*. This species is strictly a native of the near East, but arrived in Britain inside imported galls in the 1830's. The galls, which contain up to 17 per cent of tannic acid, were extensively used in the dyeing and ink trades.

Oak galls were one of the major plant ingredients used in early inks, as they contain not only their own dark pigments, but a great deal of tannin, which fixed the ink on paper and stopped it fading. Other tannin-rich barks were also used, particularly those of blackthorn, alder and dogwood. Dorothy Hartley has described how one fraud, at least, was unmasked because of an injudicious use of dogwood:

The fraud of the fabulous Baron of Arizona who claimed the vast Peralta properties, hoodwinking the U.S. Government and enjoying a millionaire's life for years, was discovered by chemical analysis proving the original fifteenth century MS. was written with iron ink, and the forgeries of 1880 were written with dogwood ink.

Food in England, Macdonald, 1954

These inks are all black. If you wish to make a red ink, fugitive but bright, steep the petals of corn poppies in a small amount of hot water and alcohol.

Lichens as litmus

I touched briefly on page 24 on some of the striking colour changes that occur in lichen extracts when they are treated with inorganic chemicals. One of these is so decisive that it is used as a standard test for acids and alkalis. Various species of *Roccella* (mainly from Africa) can be made to yield a blue dye that turns bright red in contact with acid solutions. The dye, known as litmus, is an indispensable preparation in every laboratory.

But lichens are also sensitive to the chemicals around them whilst they are still alive; and it is beginning to look as if the species that grow – or fail to grow – in our cities and countryside will have as important a role to play as indicators of the acidic substances in the atmosphere as their inert extracts do.

Botanists first began to notice how sensitive lichens were to polluted air more than a hundred years ago. Grindon attributed their decline round Manchester to the increasingly grimy air as early as 1859. Seven years later, Nylander reached the same conclusion after studying the lichens round Paris, and went on to say, prophetically, that they 'donnent à leur manière la mesure de salubrité de l'air'.

Almost every piece of research work done in the field and

in plant laboratories has backed up these observations. Lichens are more vulnerable to pollution in the air than any other type of plant. Many species are so sensitive that they simply cannot survive in the neighbourhood of towns and cities. Others have their growth stunted, or their ability to reproduce impaired. The chief culprit in all this has been identified as sulphur dioxide (produced in enormous quantities by the burning of coal in power stations and smelting works), and many of the ways in which it actually interferes with the growth of lichens have been tracked down. So exactly do these effects correlate with the concentration of sulphur dioxide in the air that it has been possible to construct a pollution index based on the kinds of lichen that are able to grow in a given area. In its fullest form this is a ten-point scale which can measure average sulphur dioxide levels in the air with an error of not much more than 10 per cent. More crudely it lays down that the more lichen species that appear in an area, the more these include large, leafy and bushy species, and the higher all of them extend up the trunks of trees, the cleaner the air is in that place.

Why are lichens so susceptible to air pollution? To answer this we need to look more closely at the ways lichens live. They are not one of our best known groups of plants, and at the rate they are being driven out of the built-up areas, they are likely to become even less familiar. But a few species are very common indeed. Those crusty grey-green patches on city trees, those brown patches on concrete walls, and those orange whorls on roofs and grave-stones, are all lichens. But the whole group is much more varied than this in both form and habitat. There are bushy lichens which hang like mistletoe from the tops of trees, and scaly species that grow in flat rosettes on seashore rocks. But they all have one important feature in common. Lichens are not a simple, single plant, but two, living in a close and mutually beneficial partnership called symbiosis. One of the plants is a fungus, which cannot manufacture its own food (see p. 152) but can provide a solid 'shell' and support. The other is a single-celled alga which can only thrive by itself in very specialised conditions, but which, being a green plant, can manufacture food by photosynthesis. Living together in a lichen cooperative, the alga has physical security and the fungus a source of nourishment.

But lichens – even those that live on the ground – lack true roots (though they have fibrous holdfasts with which they clamp themselves to flat surfaces). For minerals and other food substances which the alga cannot manufacture, they have to rely almost completely on nutrients washed their way by the rain. They have evolved very efficient means of taking in these nutrients. They can, for instance, absorb dissolved minerals over their whole surface, even from damp air. In one experiment, living specimens of a beard lichen suspended in a heavily polluted area accumulated *six times* as much sulphur from the air as cotton wool or dead lichen.

It is this absolute reliance on rainborne nutrients which may be the reason lichens are so much more sensitive to air pollution than other plants. Sulphurous acid (which is what sulphur dioxide dissolves to form in water) is toxic to most green plants. It slows down photosynthesis, poisons reproductive cells, and chokes up a plant's 'breathing' mechanisms as much as it does a lung's. But higher plants have ways of coping with quite high levels of pollution. Green leaves have pores, 'stomata', which close up at night. Lichens' pores have no such safety-valve system, and are open the whole time. Most green plants, moreover, shed their leaves once a year together with any poisons accumulated in them. Lichens live and grow continuously, making use of whatever moisture and nourishment comes their way, regardless of the season, and have no efficient ways

PLATE 7. *Woods*

1. Ash, *Fraxinus excelsior.* Parts used: young branches, mature wood, inner bark.

2. Elder, *Sambucus nigra:* young pith filled branches and mature heartwood.

3. Silver birch, *Betula pendula:* wood, twigs and bark.

4. Elm, *Ulmus procera:* mature wood and twigs.

8

of getting rid of toxins. (It is significant that in areas heavily polluted by sulphur dioxide, evergreens often start behaving like deciduous trees, and shed their contaminated leaves in the autumn.)

But lichen species vary in their sensitivity, depending on their chemical and physical make-up, and this is why it is possible to use the presence or absence of species as an index of pollution levels. The largest and leafiest species are the most vulnerable, perhaps because they are more exposed to direct rainfall, and have a higher surface area to weight ratio. The flattest are normally slower growing, less exposed, and in many cases don't absorb moisture as readily as the bushier species. But much still needs to be uncovered about the way that different lichen species respond to air pollution.

Still, the empirical relationships are quite clear, and give us an invaluable air pollution early warning system. For fuller accounts of using lichens in this way, I would refer readers to B. W. Ferry's definitive book *Air Pollution and Lichens*, or my own short layman's account in *The Pollution Handbook* (both in the bibliography). But in essence, this is how you can obtain a rule of thumb idea of the air pollution in a given area: (the lichen types are illustrated on plate 1)

a) Where there are no lichens at all, but just dusty green algae on trees and stone, sulphur dioxide pollution is very high indeed.

b) Things are a little better – but not much – where there are only crusty *Lecanora* species near the bottom of trees and on walls.

PLATE 8. *Woods*

1. Sycamore, *Acer pseudoplatanus*. Parts used: mature wood.

2. Beech, *Fagus sylvatica:* young branches and mature wood.

3. Hazel, *Corylus avellana:* branches, twigs and bark.

4. Crack willow, *Salix fragilis:* young branches and twigs.

c) Where lobed and leafy lichens begin the air is moderately clean.

d) The purest air allows the growth of bushy species, and especially beard-like lichens of the *Usnea* family.

When you are searching for indicator lichens make sure you examine a wide range of habitats in your chosen area: trees, stones, old walls, roofs, etc. And do not place too much significance on a single survey of a limited area. It may be a region of very low rainfall. The trees and buildings may be too young to have built up good populations of these slow-growing plants. Long-term change is the chief thing to be alert to – particularly the ominous disappearance of lichens from sites (otherwise unchanged) where they used to thrive.

There are two features of this approach which seem to me worth emphasising. First, it provides a way in which a large number of amateur naturalists can keep a watch on the pollution in their community without the need for expensive equipment or specialised knowledge. Secondly, as a technique which involves *living* indicators, it is a more reliable measure of average levels of pollution over a long period than chemical analyses of random samples of air. But the obvious comparison with that other 'living indicator' – the miner's canary – is not only unkind but inaccurate. These unfortunate birds respond purely to the gases present at the moment they are thrust forward. Lichens have to live *permanently* in the atmosphere we create. The pattern of their growth is not a snap judgement, perhaps on an untypically 'off' day, but what one could literally call a measured response (much as chronic bronchitis is in human indicators.)

Which is to say very little about these innocent victims of our unsavoury habits. It would be a mean tribute indeed to say that the only useful lichen was a dying one. In fact they are an extraordinarily versatile group of plants, and probably one of the most underexplored. We have seen already how important they have been as sources of natural dyes. Because of their absorbent qualities, they are also used in perfumery as fixatives (see p. 28). And recently many species have been found to contain potent antibiotics, particularly against the 'gram-positive' bacteria that cause skin infections (see illustration below). These last two properties must have been discovered very anciently, for lichens were one of the major ingredients of the stuffings of Egyptian mummies. They were mixed with sawdust and spices, and, packed into the body cavities, presumably acted as deodorants and helped slow down decay. (This practice

The lichens *Cladonia impexa* (left), and *C. furcata*, both common on heaths and moorland. Cladonia species contain didymic acid, one of the most powerful inhibitors of the tuberculosis bacillus.

led, incidentally, to one of the most gruesome of the old herbalists' sympathetic drugs – the compacted mass of dried blood and lichen scraped from inside the skulls of embalmed corpses. Known as 'mummy' it was believed to be an infallible remedy for bleeding and 'putrefaction' as late as the 18th century.)

But it is perhaps as landscape plants that lichens are most socially useful, as mellowers of the bleak surfaces of new stonework. Ironically, these are the places they are least likely to get a hold unless we make a radical improvement in the cleanliness of our air.

City lichens seem to me the most poignantly symbolic of

Architects and planners often add insult to the considerable injuries they inflict upon urban lichen populations by using glycerin-soaked sprigs to represent trees in their models. It is pleasing, then, that a cross-section of a lichen thallus should have been chosen by one designer as a quilt pattern.

all plants: clinging precariously to our monolithic structures, without even a root to give them independence from the poisons we shower upon them.

Marshmallows

Not, in this case, those succulent but teeth-corroding sweets, but custodians of the teeth made from precisely the same vegetable base. The roots of marshmallow, *Althaea officinalis*, have long been used in medicine for the soothing, gummy substances they contain. In France the dried roots (*Hochets de Guimauve*) are sold in chemists as teethers. They are hard and fibrous enough for a baby to chew on, but slowly soften on the outside as their mucilage is released (which, as a bonus, has a slight calming effect on the stomach lining).

Marshmallow roots also have the advantage of growing ready shaped for the job. They resemble thin, pale yellow carrots and can be held and sucked just like a dummy. The plant – velvety in both grey leaf and pink flower – grows amongst the higher and more sheltered corners of salt-marshes. It is still quite common in France and Germany, but less so in England, perhaps because it was once plundered and sold by the shadier London apothecaries as an infallible poultice for damage to just about every human tissue. (Culpeper lists wheezings, chin-coughs, swellings, pain or ache in the muscles, 'morphew . . . bloody flux . . . sharp fretting humours . . . ruptures, cramps or convulsions of the sinews . . . the impostumes of the throat commonly called the king's evil, and of those kernels that rise behind the ears, and inflammation and swellings in women's breast' amongst the disorders it will cure.)

Many other roots and twigs have been used for massaging gums and teeth, particularly in Africa and the Caribbean. In North America horseradish and peeled dogwood shoots were popular. So were the roots of alfalfa, which, if peeled and hammered flat, will fray into the physical form of a toothbrush.

English adults have traditionally been less particular about their teeth (though John Evelyn recommends beech-buds 'Winterhardned and dried upon the twiggs' as tooth-picks); but Dorothy Hartley quotes a most extraordinary 17th century 'Children's Necklace for the Teeth', which must have given harassed mums many hours of untroubled leisure, containing, as it did, one of our most potent nar-cotic plants (henbane), liberally soaked in alcohol.

'Take roots of henbane, of orpin and vervain and scrape them very clean with a sharp knife, cut them in long beads and string them green, first henbane, then orpin, then vervain, till it is the bigness of the child's neck. Then take as much red wine as you think the necklace will take up and put thereto a dram of red coral, and as much single peony root, finely powdered. Soak the beads in this for twenty-eight hours, and rub the powder on the beads, and when red and dry, let the child wear them.'

Quoted in *Food in England*, Macdonald, 1954

Nut polishes

The kernel of a nut is a ready-made polisher. It consists of a high proportion of oil, held and evenly distributed in a solid but friable base. When a nut is rubbed over a piece of raw wood the fibres break up, the oils are released and particles of cellulose become caught in the surface grain structure, adding a kind of satiny finish. The one disadvantage of using a nut in this way is that it is rather fiddlesome to hold, and you end up grinding your fingertips into the wood as well.

To give you some idea of the polishing power of a single nut, one good-sized hazel is usually enough for a thorough oiling of a walking stick. On new wood you will get the best results if you work several nuts into the surface, leaving a few hours between each application, and rubbing down with a duster in between.

Almost any species of nut – almond, hickory, hazel, beech – will work, each one giving a slightly different fragrance and finish. You should experiment to see which you prefer and find most convenient to work with. But there is little doubt that the best of all are walnuts. They are the biggest nuts of the region, with a light, pleasant aroma, and if picked in the autumn can contain up to 60 per cent of fatty and oily matter. In North America, wild walnuts are comparatively easy to find, as the black walnut, *Juglans nigra*, is native in woods across most of the temperate states. But in Britain and north Europe they are scarcer. We have no native walnut, and the species most often found, *Juglans regia*, is an introduction from south-eastern Europe which has only become naturalised in the warmest areas. And in Britain the frequency of late frosts during the walnut's flowering period means that even planted specimens in woods and hedgerows often do not set nuts. Not that there are many of these trees left. *Juglans regia* has one of the most elastic and beautifully figured timbers of all, and between the 17th and 19th centuries enormous numbers were felled for gunstocks and furniture. John Evelyn made an impassioned case for their replanting in *Sylva* (1664) drawing up such a formidable list of the tree's 'vertues' that it is surprising his plea was ignored.

. . . some wood especially, as that which we have from Bologne and New-England, very black of Colour, and so admirably streaked, as to represent natural flowers, Landskips, and other Fancys: To render this the better coloured, Joyners put the boards into an Oven after the batch is forth, or lay them in a warm Stable, and when they work it, polish it over with its own oyl very hot, which makes it look black and sleek . . . Besides the uses of the Wood, the fruit with husk and all when tender and very young, is for preserves, for food, and Oyl, of extraordinary use with the Painter, in whites, and other delicate Colours, also for Gold-size and Vernish; and with this they polish Walking-staves, and other works which are wrought in with burning; For Food they Fry with it in some places, and use it to burn in Lamps . . . and the very husks and leaves, being macerated in warm Water, and that Liquor poured on the Carpet of Walks, and Bowling-greens, does infallibly kill the Worms, without endangering the grass; not to mention the Dye which is made of this Lixive, to colour Wooll, Woods, and Hair, as of old they us'd it.

No wonder that Evelyn was able to report that in Germany (where a mature tree can still yield up to 100 kilos of nuts a year), 'no young Farmer whatsoever is permitted to Marry a Wife, till he bring proof that he hath planted, and is a Father of such a stated number of Walnut-trees, as the Law inviolably observed to this day, for the extraordinary benefit which this Tree affords the inhabitants.'

A walnut 'walking-stave' shined up with oily walnuts

would be a prize indeed, as elegantly self-contained a delicacy as squids cooked in their own ink. But unless you have
a mature and plentifully branched tree to crop, it is likely to
remain a dream. For a more accessible, albeit cruder, alternative try a hazel stick rubbed down with cobnuts.

Nuts are not the only plant products that have been used
in polishes. The seeds of sweet cicely, *Myrrhis odorata*, con-

The seeds of sweet cicely.

tain a small amount of oil that has the same fresh aniseed
fragrance as the leaves. Sweet cicely is a feathery umbellifer
from southern Europe, but has become naturalised in
grassy places and waysides as far north as Scotland. The
long, cylindrical seeds begin to mature in June, and in
Westmorland this is when they were warmed and pounded
with tallow to make a scented polish for oak. But if the
seeds are picked young and green enough they can be used
as they stand, like nuts, without either duster or additional
wax. They leave a very slight green stain on new wood, but
this vanishes after a few rub-downs.

Another polishing plant which gives a satiny finish shot
through initially with green is the leaf of *Philadelphus
coronarius*, sometimes known as mock orange, but more
usually, and confusingly, as Syringa (which is the scientific name for the lilacs). Mock orange is a shrub native

to southern Europe, much planted in gardens and now becoming naturalised in waste places. The leaves are very mucilaginous, and they were apparently used by housewives in 19th century Wales for shining up old furniture. They are quite effective as self-digesting dusters – provided you follow them up with a real cloth.

Paper

Our word paper derives from papyrus, the writing material made in Ancient Egypt by sticking together strips cut from the pith of the paper reed, *Cyperus papyrus* (probably Moses' 'bullrushes'). No close relative grows in the West, and I've been unable to trace any similar techniques with alternative plants. But there is a tradition of writing directly onto barks, particularly those from the birch family. One north American species with an almost pure white bark is actually known as the paper birch. It has been used not only as a writing material, but as a skin for canoes and a covering for roofs. It is that impervious to water. In fact the barks from all birch species share a remarkable resistance to wetting and decay, which has meant that we have tangible evidence of their ancient use as paper. Only recently a number of 16th century Russian birch bark scrolls were discovered in a moderately legible state. (It's interesting that, although conventional pulp paper had been in use for some time by this period, the more durable birch bark was still preferred for documents like family records, where long life was imperative.)

The manufacture of paper by the pulping of plant fibres is a later development, and was too technical a process ever to become a widespread domestic craft. It is possible to make paper at home of course, but it is a complicated, messy business, and exceedingly trying on the patience. For those readers who feel that the quality of the result is worth a certain amount of effort, I would recommend specialist books like John Mason's *Paper Making as an Artistic Craft* or the shorter note in *Country Bazaar* (see bibliography). I will confine myself here to the barest details, so that you are at least aware of the principles (and complications) of the job.

Almost any fibrous leaves or stems can be used. Nettles, cow parsley, and wild flags are particularly good; though every plant produces a paper of a unique texture and colour. In essence the process consists of separating the plant's fibres from its more fleshy parts, bleaching and pulping them, and spreading the mash as a thin paste over a flat mould to set.

The plants are stored until they begin to rot, are boiled and pounded in caustic soda, and the fibrous remains soaked in bleach until they are a pale biscuit colour. The fibres are then washed, wrung dry, shredded with shears and crushed in a pestle and mortar. The resulting pulp is transferred to a tank of warm water, and scooped up into a kind of flat sieve made from a rectangular wooden frame and a sheet of perforated zinc. The pulp is distributed as evenly as possible over the zinc, allowed to drain, and then turned out onto a wet blanket. Another blanket on top, plus a heavy weight, completes the process. And that is a sheet of paper; only one, but at least your own.

Pewterwort

Members of the horsetail family have the curious property of absorbing large quantities of silica (the basic ingredient of sand) from the soil and redepositing it as fine crystals on their stems and leaves. You will need a microscope to see these clearly, as translucent flinty grains ranged irregularly along the plant's outer skin. Yet they are substantial enough to make a horsetail feel like fine sandpaper if you run it through your hand.

Horsetails have been used anciently as vegetable files and scourers. Dairymaids polished up their milk pails with them. Medieval European fletchers and archers (and North American Indians) used them to smooth the shafts of their arrows, thus sharing at least one accoutrement of battle with their knightly commanders, who had their suits of armour shined up with these 'scouring rushes'. John Aubrey reports that they were even used in watchmakers' workshops, for giving an extra smoothness after filing.

But it was as an ordinary household scourer of pots and pans, a precursor of our wire wool, that horsetail earned its common name of pewterwort. The most effective species is *Equisetum hyemale*, a stiff and hefty specimen often growing over a metre tall. This is native in damp places in North America and Europe, but very rare in Britain. Most supplies for Britain were imported from Holland (hence the name 'Dutch rush'). But it has been a commercial vegetable in many places in the Continent, and still appears occasionally on market stalls in Austria and Eastern Europe.

But the common horsetail, *E. arvense*, is not a bad substitute, and a few shoots will serve very well for cleaning plates and pans at a picnic. Although the whole plant carries the silica crystals, they are most concentrated – and most strongly supported – on the stems, so bunch these together to make an effective brush. They leave a slight green stain on metal, but this will rinse off easily with plain water.

You should have little trouble in finding common horsetails. They grow abundantly on bare, disturbed ground, particularly if it is poorly drained. And between May and October they are unmistakable with their jointed stems and whorls of brittle, bristly fronds. They look less like horses' tails, really, than bottle brushes, and this is another local name for the plant, though not a practical use. But they are directly descended from the giant horsetails that flourished in the Carboniferous era 300 million years ago, and this gives a clue to their odd appearance and the source of all that silica.

Horsetails are of such ancient ancestry that they are in a botanical class all of their own, somewhere between the ferns and the firs. They reproduce partly by spores, which are usually carried on separate shoots like brown asparagus spears in early spring, and partly by a complex underground system of roots and tubers. This is so extensive and deep compared to the size of the overground shoots, that it is able to filch (and concentrate) silica and other minerals from an enormous volume of soil.

Incidentally, if you would like a vegetable metal polish to use with your pewterworts, try the acid juice of cranberries which has long been used for cleaning silver in Scandinavia.

Razorstrops

Razorstrop fungus, *Piptoporus betulinus*, is a shell-shaped bracket fungus which grows exclusively on birch trees. When young its flesh and undersurfaces are so smoothly rounded and white that it looks as if a slightly squashed puffball has been stuck to the trunk. As it ages, the bracket expands and flattens, and the upper surface turns a flaky grey-brown. If you cut open a young specimen you will see that the flesh is pure white, moist and almost rubbery in texture. But as it dries it becomes corky, and it was in this condition that strips of the fungus were used as razorstrops. Charles Badham, one of the first intelligent writers on fungi, described how a related species *(Polyporus squamosus)* was made up into strops. It was cut from the tree

in autumn, when its juices have been dried and its substance has become consolidated; it is then to be flattened out for twenty-four hours in a press, after which it should be carefully rubbed with pumice, sliced longitudinally, and every slip that is free from the erosions of insects be glued upon a wooden stretcher. Cesalpinus knew all this! and the barbers in his time knew it too; and it is not a little remarkable that so useful an invention should, in an age of puffing, advertisement, and improvement, like our own, have been entirely lost sight of.

On the Esculent Funguses of England, 1863

(Though those early Italians may not have been strangers to 'puffing and improvement': Badham relates how one recommended the use of surplus razorstrop strips as 'an excellent detergent, with which to brush and comb out the scurf from the hair.' A case, I suspect, of excess being the mother of invention.)

There's not much call for razorstrops these days, but the dried flesh of *P. betulinus* is so lightweight and pliant that it can be used as a substitute for cork. Under the name of *Polyporus*, it is sold to entomologists as a mounting base for small insects. It is not an end which I would wish on either insect or fungus; but if you think you have a use for a cork substitute, don't be put off by Badham's unnecessarily complicated instructions. You need do no more than cut the fungus from a dead birch, slice it into pieces about half an inch thick, and leave them to dry in a warm, damp-free room. Depending on the age of the fungus when you picked it, the strips will dry in something between one and four weeks' time. And since specimens can grow up to 30 cms across, you can, by slicing them *horizontally* (use a saw or carving knife to do the rough cutting if necessary), cut out quite substantial corky boards, for which there can be any number of uses in kitchen, office or nursery. They will even take water-colours, if you wish to decorate them. A friend of mine has produced some quite exquisite miniature plant paintings on the dried slices.

I should add that the dried fungus is pure white, quite odourless and completely non-poisonous.

Reeds and rushes

Watch a reedbed shifting in the wind and you will see why the reed and its many relatives have been so valuable as raw materials throughout human history. The tall stems give with the wind, and seem to be quite impervious to all but the fiercest storms and rains. But sometimes they will plait of their own accord, and you can see in nature the precursor of the weaves that have been one of the most important economic uses of this class of plant. One piece of imaginative reedcraft, a wild duck trap known as a 'fowler's pipe' is actually constructed amongst growing reeds.

It is the common characteristics of the stems of reeds, rushes and sedges that make them so useful and adaptable. They are long and straight, always lightweight and often hollow. The grouping of tough fibres round the outside of the stems makes them pliant, durable and easy to work. They are ideal, therefore, for weaving or bunching into hardwearing articles like baskets or mats. So many species share these qualities that in some ways we are talking about a single material. It is possible to make a corn dolly with dried rushes, or rush-mats out of couch grass. The species which has become traditionally established as the best material for a local product depends as much on its abundance in that area as on its suitability. So, for thatching in Britain, for instance, marram-grass (a sand dune species) is used in the coastal areas of Angelsey, wheat straw in the arable counties of southern England, and reed in the damp fens of East Anglia.

The common reed itself, *Phragmites australis*, is one of the most adaptable of all grassy plants. It is abundant at the edges of still water throughout the temperate zone, sometimes growing over 3 metres tall, and forming extensive beds by means of its creeping roots. Reed flowers in dense plumes of purple from August onwards, and is cut in the winter when the leaves have died away.

Reed is generally regarded as the best material of all for thatching. It is straighter and more durable than wheat straw, and the smoothly rounded, hollow stems cause rainwater to drain down the thatch rather than through it. Thatching houses is a highly skilled craft, and not something to be undertaken lightly at a do-it-yourself level. But the concept behind it – the laying of tied bundles of reed over a surface to deflect the rain – is not difficult to translate into small scale work. Try making a thatched roof for your bird table, using a simple slatted wooden frame, and tying

A bird-table roof – a modest exercise in thatching.

small bundles of dried reeds in layers over it. I have even seen lightly thatched tea-cosies in craft shops – which is hardly the most functional use, but does produce a very pleasing effect, and must be an easy way of getting to grips with the qualities of reed.

Reed has also been used as a foundation for hand-made

bricks, for making whistles, and occasionally for weaving into mats. But it is one of the less easy grass-like plants to work into curves (it can splinter, even when damp) and the best materials for weaving are to be found amongst the rush family. Rushes are structurally similar to reeds and other grasses, though their long cylindrical stems (which are sometimes filled with white pith) are generally more leathery and flexible. The most popular rush for weaving into mats and baskets is the true bulrush, *Scirpus lacustris*, which occurs in colonies at the edges of rivers and lakes all over the temperate zone. It is a formidable plant, with soft rounded stems which can grow up to 3 metres tall and as thick as 2 cms at the base. They are straight and jointless, which makes them ideal for weaving. They should be cut in late June or July, whilst the fuzzy red-brown flower-heads are still in full bloom. Later in the summer the stems tend to become too woody to work easily. The rushes should be stacked in bundles to dry for a few weeks, preferably protected from strong light as well as from the damp, as this can bleach them.

Before use, the rushes must be moistened in order to make them pliable enough to weave. This is best done by laying them on the ground or in a bath, and then dousing them from a watering can. Keep turning the rushes, and allow the surplus water to drain away. Wrap them up tightly in a damp cloth and they will be ready to use in about four hours. The basic units for simple rushwork are the plait and the bundle. For plaits, three rushes are woven together exactly as you would braid hair or rope. For bundles, a handful of a dozen or so rushes are bunched and tied together securely every 5 cms with waxed thread. The plaits or bundles can then be sewn or woven into finished articles – shopping and work baskets, wine bottle covers, lampshades, seat bottoms, table and floor mats. In Nevada, the Paiute Indians weave rush cradleboards for their babies and, with a real understanding of the natural waterproofing of a plant that spends its life up to its knees in the wet, makeshift boats for fishing and wildfowling. (They also use the stems and leaves of another reedy plant, the reedmace or cattail, *Typha latifolia*, in boatmaking.)

Hopi Indian rush-work

Because rushes are physically adapted to surviving in the
damp, there is no need to varnish articles made from them.
In fact varnish (and to a lesser extent paint) can make the
rushes brittle and therefore less hardwearing. Rushes have
a marvellously subtle coloration in their own right, especially when they are young and still mottled with pastel
greens and pinks and fawns. They do bleach eventually to a
uniform straw colour, but this too can be very attractive if it
is kept clean, so don't be afraid of *washing* your rushwork.

The most strongly weather and water resistant grass is
probably marram, *Ammophila arenaria*. Across much of
Europe this wiry, tufted perennial is one of the first plants to
colonise loose sand dunes. Its sharp, broad, pale-green
leaves are very distinctive against the sand, and are perfectly
adapted to the harsh conditions of seashore life. They are
stiff and springy and can stand up to the most violent winds.
The outside of the leaf is covered with a thick glossy cuticle,
which protects it from the abrasive power of blowing sand.
The inside surface is covered with fine hairs and grooves
which help conserve water. In drought conditions, or when
the wind is particularly drying, the whole leaf rolls into a
tube round this inner surface to enclose a layer of moisture,
and to reduce the area of leaf surface exposed to the wind.

This resilience lends a degree of indestructibility to articles made from marram grass. Dorothy Hartley has written

of an ancient marram stool she found underneath a layer of skulls in a derelict Irish church:

Reasoning from the date when the church alterations were made, I reckoned that specimen of grass must have been at least 80, and more probably, 120 years old. It was woven in a circular form and had been filled with peat moss, or some similar substance, and sewn up to form a hassock or stool. By its position in the depths of the ruined crypt, under stones, dust, and old bones, it had probably been used as a basket to carry down rubbish when the place was cleared. It must have been damped and dried repeatedly, and then left forgotten, yet the grass was as elastic and firm as the grass I had seen growing on the dunes at Newborough, across the water. Under such conditions straw, reed, or rush must have perished, but the marram grass endured.

Made in England, 1939

But the most apt use ever found for this denizen of the sands must surely be the pair of marram beach shoes which Herbert Edlin noticed in a shop in Dorset!

Marram grass leaves were harvested in late summer, and were cut well below the surface of the sand in order to obtain the longest possible blades. They were stored and dried until the winter, when they were plaited into the strips from which mats and baskets were later sewn. But the large-scale cutting of marram is no longer encouraged, because of the vital role it plays in stabilising dune-systems. It is not just the leaves which are tolerant of salt-gales and blown sand. The root system is an immense and convoluted mat of thin fibres which binds the sand down. Even when a marram plant is buried under feet of sand after a 'blow-out', new leaves immediately begin to sprout both from roots and the buried stems. Without marram grass, trapping sand above ground and binding it below, the dunes would soon be blown and scattered miles inland. In a few areas of Western Britain the harvesting of marram grass for thatch and plaiting was having such a serious effect on the stability of the coastal dunes that it was made illegal.

Dorothy Hartley believes that it was the natural pattern of woven marram and rush that gave rise to the character-istic curves and scroll-work of Celtic design. Straw is cylin-drical and hollow, and when bent will always flatten and crack into sharp angles. But both the flat blades of marram grass, and the pith-filled stems of rushes, can be eased into smooth and complex curves. Up until the 18th century thin stems of common species like the soft rush (*Juncus effusus* – see p.134) and compact rush (*J. conglomeratus*) were plaited into dolls and toy cages and other 'fancies'. Even the abundant grasses of grazing meadows can be used for some kinds of weaving, as William Cobbett demonstrated in his spirited writings on the making of straw hats. During the 1820s Cobbett became dismayed by the way the traditional English cottage craft of bonnet-weaving was being put out of business by the importation of far superior hats from Italy. It was widely believed that the fine colour and durability of these Italian hats was due to the Mediterranean sun; or to the use of some species of grass that did not grow in Britain. Either way there was little that could be done about it. But Cobbett regarded such pessimistic assumptions as an affront to his principles of cottage economy, and he was able to prove that the durable qualities of the Italian bonnets were a result not of the species of straw used, but of the time it was cut; and that their honey-blonde colour was not a product of foreign sunshine, but of a special bleaching process. The straw for English hats had traditionally been cut when fully ripe, when it was coarse and brittle and in fact already decaying. As a result it was prone to fraying and fading. The Italian straw was cut whilst the grass was still thin and green. The bleached effect was obtained by soaking the straws in boiling water for about ten minutes, and then strewing them thinly across a lawn or yard for a week to soak up the sun. Under this treatment, Cobbett found that abundant wild grasses like crested dogstail, *Cynosurus cristatus*, rye grass, *Lolium perenne*, sweet vernal grass, *Anthoxanthum odoratum* (what bonnets this must have made, scented with new-mown hay whenever the air was damp!), and especially yellow oat-grass, *Trisetum flavescens*, produced straw that was fine and durable and of a colour 'as beautiful as it possibly can be.' But he urges experiments with other wild grass species:

The flowers of wild oat, *Avena fatua*, which were once used as cheap alternatives to flies by trout fishermen.

. . . the women in labourer's, farmer's, country tradesmen's, and even in gentlemen's houses, will make collections for themselves. They will find out, I warrant them, those sorts of grass, which, when turned into bonnets, make a face look prettiest. There will not remain many banks and hedges in the kingdom unexplored for the purpose of discovering grass wherewith to make rare and beautiful straw.

Cottage Economy, 1823

Rushlights

Elsewhere in this splendid tract on the virtues of self-sufficiency, *Cottage Economy*, William Cobbett has great praise for rushlights. Their chapter is sandwiched between goats and mustard and Cobbett obviously regarded them as one of the staples of life:

I was bred and brought up mostly by *Rush-light*, and I do not find that I see less clearly than other people. Candles certainly were not much used in English labourer's dwellings in the days when they had meat dinners and Sunday coats. Potatoes and taxed candles seem to have grown into fashion together; and, perhaps, for this reason: that, when the pot ceased to afford *grease* for the rushes, the potato-gorger was compelled to go to the chandler's shop for light to swallow the potatoes by, else he might have devoured peelings and all.

Cottage Economy, 1823

'Rushlight' is almost all that is needed as a description of these ancient vegetable tapers: they are lengths of peeled rush, soaked in some inflammable substance and burned (usually singly) like disembodied wicks. Structurally, rushes are ideal plants for the job. They are tall and cylindrical, have a stiff, supportive stem and often contain a pith which will absorb grease. Any sizeable kind with a continuous pith will do to make the lights, but the species most often used is the soft rush, *Juncus effusus*. This is common in badly drained ground, growing in tufts over a metre high, and has smooth, shiny stems of a fresh green colour that distinguishes them from almost all other

species. The loose clusters of flowers are carried a few
inches below the tops of the stems.

You should cut the rushes in early autumn, when they are
full-grown but still green. (They are almost impossible to
peel once they have turned brown.) Cut off both ends and
keep the thick central portion, which may be anything
between 20 and 60 cm long.

The only difficult stage in making a rushlight is peeling
the outer skin away from the pith. Strips tend to drag
chunks of pith with them as you draw them off, or taper
away to nothing as frustratingly as underripe plum skins.
But you can take steps to make the peeling less awkward.
Try, firstly, to peel your rushes as soon as you have picked
them, before the skin starts to dry. Practise on 20 cm
lengths until you have the knack (even these short tapers
can burn for up to a quarter of an hour). And use the point
of a knife to divide the skin at one end, so that there is
something for your fingernails to hold on to.

Carry on removing the peel until there is just one thin
strip about 1 mm wide the whole length of the rush, to act as
a spinal support for the pith. (Try and have a close-look at
the pith at this stage through a lens. You will see the fine
cellular structure which gives it its absorbent properties.)

Almost any fatty material will do to impregnate the
rushes: tallow, lard, and bacon fat have all been used in the
past. Dripping from a joint of roast meat is one of the most
economical fats to use, and produces good sturdy tapers.
You need do nothing more than melt the fat in a wide pan,
and soak the rushes in it for about 30 seconds (or run them
slowly through if they are too long for the pan). Let the
excess fat drain from them, and then rest them somewhere
to cool and set. In Cobbett's day they were laid in a trough
of bark suspended by straps from the wall.

One great advantage of rushlights over candles is that
they don't drip scalding tallow as they burn. You can, if you
wish, carry them about without the need of any holder. For
fixed lights the rush was clipped into an iron holder. You
might be lucky enough to come across one of these in a
curio shop, but a bulldog clip balanced in the mouth of a

A rushlight holder. The rush is held diagonally at the top of the stand.

bottle will do just as well. Clip the rush into one corner of this, so that it is free of the outside of the bottle. It can be raised occasionally as it burns through. Rushlights burn with a clear, almost smokeless flame, and sometimes smell of roasting meat. They are surprisingly bright, too. A good sized rushlight would often serve several people sitting round one table, and for those doing particularly fine work, like lace-making, a globe of water would be used to throw a concentrated spotlight.

Rushlights like these have a history of use in the western world that goes back more than 2000 years. They survived the candle, and to some extent the oil lamp, but with the coming of electricity faded into oblivion. Yet during the dark days of the Second World War, when candles were in desperately short supply, they had a temporary revival in Britain. I wonder whether our recurrent energy crises might not be a good reason for thinking seriously about rushlights again. There could scarcely be a light source more appropriate for our times. They are cheap, replenishable and fuelled by recycled food waste. Their economics are as relevant now as they were in Cobbett's time, and in Gilbert White's, fifty years before.

It is White who has given us the finest account of the rushlight. It is tinged slightly with patronisation, and I doubt if that comfortably-off Hampshire vicar ever lit his own study with them; but as a justification it shows the same fine attention to detail, the same understanding of the interdependance of *all* living things, that characterised his writings on natural history:

The rushes are in best condition in the height of summer; but they may be gathered, so as to serve the purpose well, quite on to autumn. It would be needless to add that the largest and longest are best. Decayed labourers, women, and children, make it their business to procure and prepare them. . . . At first a person would find it no easy matter to divest a rush of its peel or rind, so as to leave one regular, narrow, even rib from top to bottom that may support the pith: but this, like other feats, soon becomes familiar even to children; and we have seen an old woman, stone-blind, performing this business with great dispatch, and seldom failing to strip them with the nicest regularity.

Some address is required in dipping these rushes in the scalding fat or grease, but this knack also is to be attained by practice. The careful wife of an industrious Hampshire labourer obtains all her fat for nothing; for she saves the scummings of her bacon-pot for this use; and, if the grease abounds with salt, she causes the salt to precipitate to the bottom, by setting the scummings in a warm oven. Where hogs are not much in use, and especially by the sea-side, the coarser animal oils will come very cheap. A pound of common grease may be procured for four pence; and about six pounds of grease will dip a pound of rushes; and one pound of rushes may be bought for one shilling: so that a pound of rushes, medicated and ready for use, will cost three shillings. If men that keep bees will mix a little wax with the grease, it will give it a consistency, and render it more cleanly, and make the rushes burn longer: mutton-suet would have the same effect.

A good rush, which measured in length two feet four inches and a half, being minuted, burnt only three minutes short of an hour: and a rush of still greater length has been known to burn one hour and a quarter.

These rushes give a good clear light. Watch-lights (coated with tallow), it is true, shed a dismal one, 'darkness visible'; but then the wicks of those have two ribs of the rind, or peel, to support the pith, while the wick of the dipped rush but one. The two ribs are intended to impede the progress of the flame, and make the candle last.

In a pound of dry rushes, avoirdupois, which I caused to be

weighed and numbered, we found upwards of one thousand six hundred individuals. Now suppose each of these burns, one with another, only half an hour, then a poor man will purchase eight hundred hours of light, a time exceeding thirty-three entire days, for three shillings. According to this account each rush, before dipping, costs 1/33 of a farthing, and 1/11 afterwards. Thus a poor family will enjoy 5½ hours of comfortable lighting for a farthing. An experienced old housekeeper assures me that one pound and a half of rushes completely supplies his family the year round, since working people burn no candle in the long days, because they rise and go to bed by daylight.

The Natural History of Selborne, Gilbert White, 1789

If, in the face of this impressive cost-accounting, you would still prefer real candles, it is possible to obtain a vegetable wax from the aromatic shrub, *Myrica gale,* or bog myrtle. The wax is carried in tiny yellow glands along the young twigs and leaves, and can be extracted by boiling the foliage in salted water. The wax floats to the surface like a scum, is skimmed off, remelted and strained. You will need to collect an immense quantity of myrtle to make enough

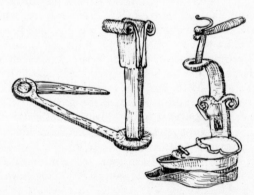

Alternative lights: a 'peer Man' (left), which held a glowing splinter of resinous wood, normally fir or birch. The 'cruisie lamp' (right), held oil and wick (which was often a rush pith) in its upper chamber, and caught drips of spare oil in its lower. Sometimes the oil used was the clear odourless and inflammable juice expressed from ripe dogwood (*Cornus* spp.) berries.

wax for even one full-blown candle, so perhaps a modestly impregnated rushlight would be a more sensible aim.

In North America candles are also made from other members of the *Myrica* family, notably the bayberry or wax myrtle. Bayberry is a medium sized shrub with a special liking for sandy shorelines. On the Pacific coast of the United States it will even grow where the spring tides lap its roots. In autumn the twigs bear clusters of greyish-white fruits. They look like berries, but this is a deception that would be worthy of a plastic flower manufacturer. In fact the 'berries' are small nutlets covered with a heavy coat of aromatic wax.

The wax is extracted in much the same way (though in rather larger quantities) as it is from bog myrtle. The berries are boiled in water for about a quarter of an hour, and the resulting mixture of molten wax and water strained through a cloth into a narrow vessel. As the liquid cools, the wax solidifies and can be scooped out and remelted.

Candles made from bayberry wax are grey-green in colour, and burn with a mildly spicy fragrance. They are sold in gift and craft shops in the Cape Cod area, an ironic reminder of a time when they were a vital source of light, not just a table decoration. And a reminder too, of just how thrifty and resourceful our ancestors were. The early settlers who made bayberry candles scarcely wasted a fragment of the raw materials. They used the water in which the berries had been boiled as a blue dye and a mouthwash, and the leaves from the twigs as a soup flavouring!

Rushmats

The use of woven rushmats (p.127) as a floor covering is really just a refinement of a much earlier custom of strewing floors with a rough layer of rushes or straw. In the days when men and animals shared eating and sleeping quarters, they also shared whatever was used to cushion the ground; and very welcome even rushes must have been when the floor beneath was bare stone or compacted earth. When Thomas a' Becket was made Archbishop of Canterbury, he ordered his hall to be strewn with sweet-scented rushes, and, with not a thought for the ill-luck it was supposed to bring, with May blossom . . . 'that such knights as the benches could not contain, might sit on the floor without dirtying their clothes'.

Plant carpets of this kind had the advantage of being disposable, and they were regularly 'mucked-out' and replaced with fresh supplies. The problem – no doubt particularly acute when eating was rather less tidy than it is today – was the smell. This was the incentive which led to the addition of scented herbs to the basic mixture of rushes. Later, fragrant carpets began to be enjoyed for their own sake. Meadowsweet was one of the most popular strewing herbs, for the sharp tang of new-mown hay and soap it gives off when crushed underfoot. Gerard was a great admirer:

The leaves and floures farre excell all other strowing herbes, for to decke up houses, to straw in chambers, halls and banqueting houses in the sommer time; for the smell thereof makes the heart merrie, delighteth the senses.

The Herball, 1633

It was Gerard who popularised in Britain the ideal strewing herb – scented leaves and resilient rush-like stems combined inside a single plant. *Acorus calamus*, sweet flag or sweet rush, is a native of Asia and north America, and arrived in Europe in 1567. In fact it is not a rush, being most closely related to cuckoo-pint and the arums. But it does grow somewhat rushily at the edges of rivers and ponds, with a tall stem carrying the banana-shaped flower spike, and long, sword-like leaves, crinkled at both edges. The leaves have a very curious smell, like tangerine peel, but with a powerful, throat-catching undertone of vanilla that some find unpleasant. But it was clearly much loved as a scent in the 17th century, and was brought into household use only a matter of years after its introduction. By 1610 it was being cultivated in the Fens. By the middle of the century it was spreading rapidly of its own accord, and was so abundant in Norfolk that Sir Thomas Browne was able to write:

This elegant plante groweth very plentifully and beareth its Jules yearly by the bankes of Norwich river, chiefly about Claxton and Surlingham, and also between Norwich and Hellsden bridge, so that I have known Heigham Church in the suburbs of Norwich strewed all over with it.

(From a letter to Merret, 1668)

By this time there was sufficient sweet flag in the Norfolk Broads to carpet the whole of Norwich Cathedral every summer, and it was probably included in the 'Rush-bearing' ceremonies that were once frequent in many country churches. In the north of England, the Church's patronage of rushes was so strong that some churches (like that in Grasmere in the Lake District) were not paved until the latter half of the 19th century.

Nowadays, in an era of fitted carpets and washable linoleum, strewing plants on the floor would seem an obstacle to hygiene and comfort, rather than a contribution. And in some ways we have incorporated the idea symbolically in the designs of our rugs, which are often in the style of formal flower-beds. But strewing is worth trying, even if only on one special occasion round a dinner table or ceremonial bed. It would hardly be defensible to gather the huge quantities of rather scarce plants that were used in the 17th century, but there are plenty of common herbs that could be used for an occasional treat. As a guideline to the species to use, choose those that behave like strewing herbs in the wild; that is, are thick on the ground and noticeably fragrant when you walk over them. Meadowsweet is as good as it was in Gerard's day, and can be found in abundance in damp meadows. Try mints, too, especially the pepperminty water mint. Salad burnet gives the characteristic smell of cucumber peel when you walk over chalk downland, and will do the same on your floor. But perhaps the most suitable of all is pineapple weed. This member of the daisy family, otherwise known as rayless chamomile, *Matricaria matricarioides*, is now abundant in waste places, and well-known for its yellow, petal-less button flowers, and the delightful smell of pineapple which its feathery leaves give off when they are crushed. It was originally a native of Asia, but over the last hundred years has spread throughout Europe and north America, particularly along tracks and roads. It is apt as a strewing plant in other ways, as its abundance by waysides is in part due to its tolerance of being trampled underfoot.

The first time I ever used aromatic plants on the floor was during a summer holiday in Spain. Our house was perched in a pine wood looking due south over the Bay of Palamos, and with a certain amount of imagination we could believe ourselves directly downwind from Africa. Inspired by those musky breezes, our cooking became increasingly coloured with Moorish extravagance. One evening we decided to cook chak-chouka, a Tunisian stew of tomatoes and eggs and chillis and hot chorizo sausages. To complete the fantasy we ate it lying on the floor, on a carpet of pine twigs cut from the trees around the house. The decadence of the occasion was as much make-believe as our proximity to Africa. But the warm, spicy resin rising from those pine needles, mingling with the smell of peaches and peppers, was decidedly and memorably real.

So do try some herbs on your floor. Throw fresh mint on your bathmat, and think of Thomas Tusser's euphonious list of strewing herbs, a sight more refreshing than the list of ingredients on the back of domestic deodorisers:

Bassell, fine and busht, sowe in May. Bawlme, set in Marche. Camamel. Costemary. Cowsleps and paggles. Daisies of all sorts. Sweet fennell. Germander. Hop, set in Februarie. Lavender. Lavender spike. Lavender cotten. Marjorom, knotted, sow or set, at the spring. Mawdelin. Peny ryall. Roses of all sorts, in January and September. Red myntes. Sage. Tansey. Violets. Winter savery.

Five Hundred Pointes of Good Husbandrie, 1573

Soapworts

No name could be less ambiguous: soapwort – a vegetable detergent. Boil the leaves (or roots) of this member of the pink family in water, and you will obtain a lathery liquid with the power to dissolve fats and grease. It even has the fresh green colour favoured by modern detergent manufacturers.

The history of soapwort, *Saponaria officinalis*, as a washing herb goes back to at least medieval times. Geoffrey Grigson has pieced together the evolution of its functional names, which probably accurately reflect its changing fortunes as an economic plant:

'William Turner invented Sopewort and Skowrwurt, because the herbarists called it *Saponaria* and *Herba Fullonum*, the 'fullers' herb'. (In French, it is still *herbe à foulon*.) The English names in *The Grete Herball* of 1526 were Burit, Herbe Phylyp (? St Philip), Saponary, Fuller's grasse, and Crowsope. Another of its medieval names was Foam Dock. Possibly soapwort was used by the early medieval fullers for soaping cloth before it went under the stamps of the mill, and no doubt it was one of the ancient washing plants before the invention or the general employment of soap. Washing plants are still used by the the Arabs, and soapwort is still cultivated for washing woollens in Syria. It was so used in France in the nineteenth century. In the Swiss Alps, sheep were washed with a mixture of the leaves and roots and water before they were shorn, and linens were washed in soapwort juice and ashes.

The Englishman's Flora, Phoenix House, 1958

Eventually, like so many economic herbs, soapwort was taken to North America by European settlers, where one of the names it acquired was My Lady's Washing Bowl.

It was about the same time that soapwort's long reign as a washhouse herb came to an end in Europe. But the plant itself still hangs on in odd patches of wasteland. It is a native of southern Europe strictly, where it grows on damp woodland edges and stream-sides. In Northern Europe, Britain and America it has long been naturalised on roadsides and railway embankments, though never far from human habitation. Some of these colonies may be genuine relics of old launderer's gardens, but increasingly they are of a double-flowered variety, escaped from the herbaceous border rather than the herb bed. The single-flowered, original soapwort has delicate, five-petalled pink flowers above its shiny, spear-shaped leaves; the double-flowered has a positive froth of petals, like a rough-cut version of its carnation cousins. One of the old country names for the plant, Bouncing Bett, seems to have settled particularly on the double-flowered variety. I must confess that I don't know Bett's pedigree. She may have been an archetypal washerwoman, or just exuberant froth. But her name suits the sudsy blooms of *S. officinalis* var. *biflora* very well.

If you have any kind of soapwort in your garden, or a plentiful source outside, do try it out. A large handful of the leaves, chopped and bruised, needs about half a pint of water. Boil them together for half an hour, then strain off the resulting soapy liquid. I have found that it cleans the saucepan in which you made it best of all, but once diluted with hot water makes rather heavy weather of a bowl of dirty dishes. It seems to work better on materials based on living products. Your hands will be well but gently scoured after struggling with the washing up, and so will any natural fabric which you boil in the solution. (It leaves a slight green stain, but this rinses out quite easily.) The detergent effect is certainly no illusion. Soapwort contains colloidal chemicals called saponins, which, like true soaps, appear to 'lubricate' and adsorb dirt particles. Vegetable saponins are less corrosive than soaps, and because of this soapwort is still occasionally used where the very gentlest cleansing is needed, for instance in the washing of ancient tapestries in museums.

Conkers, the autumn fruits of the horse chestnut, also contain a good deal of saponin (up to five per cent) as well as starch, proteins and minerals. During the Second World War, British scientists developed a process for separating out these various useful ingredients. There was nothing original in the idea of using the chemicals in conkers. They had been used for cleaning wool in France, and as a source of rough starch in Germany. But beleaguered Britain planned to use the lot, the saponins becoming foaming agents for fire extinguishers, and the final residue (which contained 18 per cent protein) a supplement to cattle fodder. In 1943, over 1000 tons of conkers were gathered by volunteers across the country and dispatched to chemical factories. Sadly, the process proved to be uneconomical on a large scale, and was discontinued. Given the soaring costs of petroleum-based detergents (and cattle feed), I think it might be an appropriate time to re-examine its economics.

A couple of final botanical laundering tips. Ivy leaves, boiled and mashed until the water is dark, make a rinse for revitalising black silk. And the juice pressed out of the leaves of wood sorrel can be used to remove iron mould stains from linen, and was once sold for this purpose under the oddly misleading name of 'Essential Salts of Lemon'.

Spindle

Spindle is a rather slender European shrub, with a special preference for woods and bushy thickets on limestone, which produces curiously square-sectioned, straight twigs. It is also, of course, a wooden spinning-stick, and it is likely that this is where the tree got its name. For thousands of years before the invention of the spinning wheel, woollen thread was hand-spun with a weighted rod called a spindle. The coarse yarn was tied to the end of the spindle, which was then set spinning so that it would draw out the thread by its own weight. The wood of the spindle tree was ideal for the purpose, being tough, heavy, smooth, and very hard. The young branches, often as thin and straight as dowels, could be cut into spindles without any elaborate carving.

But in England at least (where there was no great spinning tradition) the name spindle tree appears to have been coined by professional botanists rather than ancient usage. William Turner wrote in 1568, that 'I have sene this tree oft tymes in England and in most plentye between Ware and Barkways, yet al that I coulde never learne an English name for it, the Dutche men call it in Netherlande Spilboome, that is Spindel tree, because they used to made spindels out of it in that countrey, and me thynke it may be so wel named in English.' *(The Herbal)*

It is odd that this highly specialised name stuck, in England as well as Holland, and replaced a host of popular

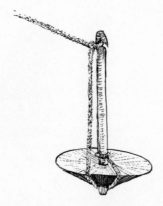

A spindle. A tuft drawn out of a hank of wool is tied to the
end of the longer shaft, and the spindle set rotating. The
spinner feeds more wool out, twisting it as it goes; and the
spindle's rotary motion provides the tension to draw the
loose fibres into a thread. The wooden disc is to give the
spindle momentum. Previously a circular stone was used.

country names. For *Euonymus europaens* was a common,
conspicuous and useful shrub long before it became a
specific for spinning-sticks. The hardness of the pale yellow
wood meant that it was used for skewers and toothpicks
('prickwood' was an early name), and later for viola bows,
virginal keys, pegs, knitting needles and even bird-cages.
But it is for its autumn berries that spindle has always been
most loved and noticed, those extraordinary fruits like
miniature, shocking pink pumpkins, whose four lobes contain – and modestly reveal as they open – perfectly round,
pure orange seeds. They are fiercely purgative, like all parts
of the tree, and probably had some use in folk medicine.
They were also (as 'louseberries') baked and powdered, and
rubbed into the hair of boys to rid them of lice.

Teasels

To 'tease' cloth is to comb it into separate fibres, or to raise the pile or downy 'nap' of finished cloth. The word's root is the Old English *taësan*, to pluck or pull. The teasel's name comes from precisely the same root. It was the plant whose spiny seedheads plucked and pulled at your clothes if you brushed past it. But I suspect that it gained its name from long service in the cloth industry rather than from any accidental combing qualities.

It is likely that teasel-heads were first used in the making of cloth for 'carding' wool, that is separating out the individual fibres prior to spinning. But murals excavated at Pompeii have shown that, by Roman times, they were also being used to raise the nap on finished cloth. During the last century they were largely superceded by steel brushes, but they are still regarded as unsurpassable in the finishing of cloths that need an exceptionally fine and even pile, like the baize coverings used for billiard tables. For specialist processes like this, no man-made substitute for teasel-heads has yet been produced, and they remain one of the very few plant products that are used as tools almost exactly in the form in which they grow.

The key to their superiority lies in the hundreds of small hooked spikes which cover the flower. Set in the head, they have a 'give', as if they were mounted in rubber. If they meet a snag or irregularity in the cloth, they will bend (or break) and skate gently over it, unlike the less flexible steel brushes, which are apt to tear through indiscriminately.

For those who cannot recall hooked spines on teasel-heads I should point out that the plant used in commerce is not the common teasel, *Dipsacus fullonum*, but fuller's teasel (fullers were those who cleaned and finished cloth), *D. fullonum* ssp. *sativus.* Although this differs in many respects from *D. fullonum*, it is generally regarded as no more than a subspecies, perhaps only a variety, whose characteristics have been preserved and exaggerated by cultivation. The most conspicuous differences are in the flower-head. The common teasel has an almost conical head with long, straight, weak spines. Fuller's teasel has a taller, cylindrical head, with stiffer, down-curved spines.

If fuller's teasel is a distinct sub-species it is probably a native of southern Europe, and in Britain crops up only occasionally in waste places (though it will self-seed in warm summers). The bulk of the teasels for use in the woollen mills are grown commercially in south Somerset. The teasel is a biennial, and the seeded flower heads are normally harvested in the August of the year following sowing, when the flowers have died off. As the teasels are cut they are bunched in groups of 30 heads, and eventually assembled in rings round tall drying staffs. These are normally left outside to dry and harden.

Once the dried heads have arrived at the cloth factory they are mounted in wooden frames, which in turn are fitted onto a rotating drum. The roll of cloth to be finished is then rotated in the opposite direction, against this drum, so that the fabric and the teasel hooks brush against one another.

Few of us are likely to be in a position to tease our own cloth, but don't dismiss the teasel as a household tool because of that. Even common teasels can do more than merely provide the centrepiece of flower arrangements. Cut whilst they still have a hint of green in them, the heads make quite serviceable clothes brushes – even hairbrushes, if you have a spartan scalp.

Tinder

The best plant materials for kindling a fire are very dry pine needles and paper-thin strips peeled from the outer bark of birch trees. They can often be made to flare up with no more than a spark from a flint or tinder-box. But true tinder is prepared from a fungus, *Fomes fomentarius*. This is a large bracket fungus, not unlike *Ganoderma applanatum* (p.153), which in Europe and America grows largely on beech trees (though in Britain it is restricted to birch woods in the Scottish Highlands). When mature it is hard, darkly coloured and shaped something like the hoof of a cart horse.

The brackets are cut from the trees in early autumn, sliced, washed with soda, and then beaten and worked to rid them of their woody fibres. At this stage the pieces resemble soft – if rather tattered – buckskin leather, and were once much in demand as a post-surgical styptic dressing. But the *F. fomentarius* harvest is now used exclusively for tinder. The leathery strips are soaked in saltpetre solution and dried. In this state they are highly inflammable, and will catch fire from the slightest spark. The thin strips can still occasionally be found (as 'Amadou') in tobacconists' shops in Northern Europe, where they are sold as cigar lighters – an odd fate for a member of a plant group usually associated with dark and damp.

Other absorbent plant materials have occasionally been soaked in saltpetre for use as tinder and wicks. The downy fibres which can be rolled from mullein were used in this way, resulting in local names like High Taper and Candlewick plant. So were those from mugwort and coltsfoot leaves. There is a curious use of mugwort tinder in the Chinese medical practice known as moxibustion. Small heaps of dried leaves are burnt on the skin until they raise a blister, on chosen points on the meridians of forces and essences which, according to Chinese philosophy, determine the body's state of health.

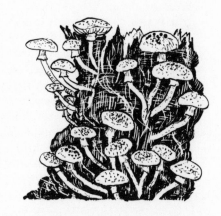

Toadstools

Of all the classes of plants, I suppose fungi are the most despised and mistrusted. They have been charged with a formidable list of offences, some based on fact, some on superstition. Eat them and they will poison you. They pull down houses and ferment jam. As late as the 19th century it was believed that they were formed from the spittle of snakes, or by the action of thunder, or from some coagulation of mud. No wonder, perhaps, that the urge to kick them to bits on a woodland floor seems almost universal. (It certainly has an ancient pedigree: Clusius, who wrote the first known monograph on fungi in 1601, when he was seventy-six, was able to recall that as a boy he had booted puffballs about.) Perhaps it is just toadstools' recumbent postures that invite such attacks. But their association with death and decay most certainly reinforces the hostility towards them. Yet it is precisely this feature that gives fungi their importance in nature, and makes any discussion of their usefulness in domestic terms of secondary importance. Fungi are one of the great groups of natural scavengers. Since they contain no chlorophyll, they are unable to manufacture their own food, and must steal it from other green plants, either living or dead. In doing so, they help break up the fibrous structures of plant remains and return their minerals and proteins to the natural cycle. Without fungi and other agents of decay like bacteria, the soil would soon be exhausted and smothered under a sterile layer of locked-up chemicals.

But if this natural usefulness overshadows all others, it remains true that fungi, with their complex chemistry and immense variety of form, have been of great economic importance. Because of their precarious dependence on other plants, fungi manufacture many complex substances to help feed and protect themselves. Some of these are enzymes for breaking down plant material into compounds which the fungus can utilise. We make use of these in brewing and cheesemaking, and in the treatment of sewage. Many fungi also manufacture poisons which are presumably useful in discouraging predators, or holding off invasions by other fungi which are competing for food. In recent years we have discovered that many of these chemicals are equally useful in fighting off human diseases. The raw material from which penicillin is derived is a fungus. So is the source of streptomycin. Ergot, the tiny, drum-stick-shaped fungus which can infect whole fields of rye and barley, is the source of two vital modern drugs: ergotamine, for migraine, and ergometrine, used to stop bleeding after childbirth.

At the other extreme, even the fleshy matter of toadstools, which can range from the gelatinously fragile to the rock-hard, has been exploited. I have discussed some of the more widespread and accessible uses of different species elsewhere in the book (see INKCAPS, RAZORSTROPS, TINDER). What follows here are brief accounts of some of the potentialities of this much misunderstood and under-explored class of plants.

*

An 'artist's fungus' sounds the most unlikely possibility of all – a joke-shop novelty, like an easel with dry rot, or an edible palette. Seeing the fungus in question (usually known just by its Latin name *Ganoderma applanatum* in Britain) does not make its title any more credible. A specimen growing down the side of a hardwood tree can look like a stream of solidified lava, rippled and moulded to the countours of the trunk. Even in texture it has the qualities of some mineral ooze. The red brown upper surface is rock-hard, but as brittle and translucent as resin. Yet it is a living fungus, and a mightily destructive one, causing a serious heart-rot in the trees it attacks. No two things could surely be more incompatible than art – or indeed any creative activity – and this malignant parasite.

But *G. applanatum*'s artistic properties lie under the surface – in the spore-bearing tubes on its underside, to be exact. These are white when young, but if bruised or scratched, rapidly stain an almost permanent violet-brown. If you break or cut a specimen in half you will see how the tubes grow in strata (another geological touch), each layer representing a year's growth, and being sandwiched between sheets of chocolate-brown flesh.

Because the staining is reasonably fast and permanent, artist's fungus can be used as a medium for etchings. A large, clean specimen with a white undersurface is cut or prised from a tree (great care being taken not to bruise the tubes) and the design scratched on the face with a needle or the tip of a sharp knife. In North America you can sometimes find specimens of this art-work in handicraft centres.

Artist's fungus is one of the more common bracket fungi, and though it can occur on any hardwood tree, it has a particular taste for beech. All the more curious, then, that its graphic potential is so little known, for the beech, whose smooth bark scars as easily as a marrow's, is the prime target in the countryside for those who want to leave their mark on the world.

Another fungus with artistic associations is the giant puffball, *Lycoperdon giganteum*, whose dried spores were once used to produce stage lightning. *L. giganteum* is frequent on pastureland throughout Europe and North

America, and in common with other species, has pure white flesh which turns gradually into a mass of olive-brown spores. The size to which the giant species can grow means that the volume of spores which are released is enormous. It is not unusual to find specimens as big as a man's head, and one this size can contain over 1,000,000,000,000 spores.

The spores, which puff naturally out of a hole on top of the dried-out skin when the puffball is fully mature, have been used for centuries to smoke out bees. Gerard wrote that:

In English Fusse bals, Pucke Fusse, and Bulfists, with which in some places in England they use to kill or smolder their Bees, when they would drive the Hives, and bereave the poore Bees of their meate, house and lives; these are also used in some places where neighbours dwell farre a sunder, to carrie and reserve fire from place to place.

The Herball

Some beekeepers still use whole dried balls as natural aerosols, and puff the spores around the hive when they wish to calm the bees. But the spores are highly inflammable, and the same effect can be obtained by igniting dried puffballs. Similarly, to produce lightning on stage, a mass of the spore dust (often mixed with spores from clubmoss) was ignited and thrown across the area where the effect was wanted. The flame flashed across the cloud of spores, much as it would across a gas. Puffballs were also valued as a styptic for small cuts, and also – though there was doubtless magic as well as medicine behind this – for burns. They were a common item at one time in blacksmiths' shops in East Anglia.

Another fungus that has been used for stupefying insects (though it is best known for its similar intoxicating effect on humans!) is the fly agaric, *Amanita muscaria*. This is the red and white spotted toadstool which appears so universally, and with such apparent innocence, in children's picture books. It is abundant in birch woods throughout the temperate zone, and has been used as a domestic flytrap since the middle ages. In the 13th century it was suggested that the cap should be broken up in milk, and jars of the mix placed around fly-infested rooms. Nowadays, in eastern Europe, the milk is replaced by sugar solution, or the caps used as they grow with sugar sprinkled directly on them.

Another curious quality possessed by some toadstools is the power of phosphorescence. The best known species in this respect is the honey fungus, *Armillaria mellea*. This tawny-brown toadstool grows in large tufts on tree stumps and roots, and is a serious parasite of living wood. It is reckoned that, in Europe at least, more trees and shrubs are killed by it than by any other parasitic agent. The honey fungus attacks new wood by means of long mycelial threads, like black bootlaces, which can be seen under the bark of infected wood. It is these mycelia which cause the luminosity of rotten wood. Pieces of infected timber, glowing a bluish-white in woods at night, were wondered at and revered before the source of the light was recognised, and it is possible that they gave rise to the idea of the magic wand. But they quickly acquired more prosaic uses, and there are records from the 17th century of blocks of rotten timber – known as 'touch-wood' – being used as way-markers in Scandinavian forests during the long winter nights. In the First World War, the troops in the trenches used to stick similar pieces of wood in the straps of their helmets to avoid collisions in the darkness, particularly when there were explosives nearby that would make naked flames dangerous.

Phosphorescence in fungi, and in other living organisms, is believed to be due to two substances known as luciferin and luciferase, which, in the presence of oxygen and water at normal temperatures, react with each other and produce light as a by-product. Given the safety, efficiency and low energy consumption of the process, we could well benefit by investigating it further.

Another fungus which can appear to give an almost luminous coloration to rotten wood is the green-staining fungus, *Chlorosplenium aeruginosum*. Wood infected by this species is tinted a bright blue-green (sometimes almost the shade of corroded copper), by a pigment called xylindeine in the mycelium of the fungus. Rotting wood scraps could never be of much economic use though, however lavishly they were coloured. But *C. aeruginosum* does grow in freshly dead wood, and the colour it gives is so attractive and unusual that pieces of it were in great demand by cabinet makers. Green oak was the most popular, and in Tonbridge, Kent, a distinctive style of inlay work was based around it. In preparing 'Tonbridge Ware' strips of differently coloured wood were glued together into a block so that their ends formed a picture or design. The block was then sawn transversely into thin strips. This mosaic veneer was finally glued to the table top, jewel or snuff box that was to be decorated. This craft has largely died out, but it did lead to a method of artificially inoculating oak with the fungus.

Another fungus-stained wood in wider use is the rich, red-brown oak produced by the beefsteak fungus. This curious bracket fungus, which so resembles a well glazed ox-tongue, grows on living wood, and so the timber coloured by it is more useful and more substantial.

But the most widely used brown-stained oak is not the product of any fungus, but of chemicals in peat bogs. Oak which has lain in the acid conditions of a bog for centuries is usually reduced to pure heartwood, and becomes stained a warm grey-brown. Much of the best bog oak is found in Ireland, where it is used to make souvenirs. But in 1948 a huge tree was dredged out of Stilton Fen in Yorkshire, and was estimated to have been 50 metres in height and 5 metres in girth. And Gilbert White, writing of the boggy country of East Hampshire, states 'I myself have seen cottages on the verge of this wild district, whose timbers consisted of a black hard wood, looking like oak, which the owners assured me they procured from the bogs by probing the soil with spits, or some such instruments.' The older villagers also assured him that they could detect the site of a buried tree by the patterning of hoar frost on a winter's day. The frost was presumed to lie longer over a concealed object than over the rest of the bog. But there is little doubt that they also knew the whereabouts of some trunks because they had deliberately buried them there to cure.

Whilst we are dealing with oak, I should mention in passing *Daedalea quercina*, a fungus which occurs most frequently on dead oak. It grows in wedge-shaped muddy-

The bracket fungus, *Daedalea quercina*.

brown brackets, sometimes as much as 30 cms across, and its underside is a labyrinth of irregularly shaped tubes. These, combined with the corky texture of the whole fungus, make it feel rather like a rough sponge if you rub it across your hand. And it was reputedly used for rubbing down the coats of thoroughbred racehorses. But I have a feeling that there was a touch of sympathetic magic as well as sensible utility in the choice of this rather scarce fungus as a stable brush; if you look at a middle-sized specimen on a trunk, it looks uncommonly like a horse's hoof.

For a final curiosity, a fungus scent-ball. The small bracket fungus, *Trametes suaveolens*, has a distinctly sweet smell of aniseed and was apparently used as a love-charm and cosmetic by Scandinavian swains. Linnaeus has an oddly snooty note about this in *Flora Lapponica* – though perhaps not so odd when you consider that, for all his genius, he was of the opinion that fungus spores were the eggs of insects!

The Lapland youth having found this Agaric, carefully preserves it in a little pocket hanging at his waist, that its grateful perfume may render him more acceptable to his favourite-one. O whimsical Venus! in other regions you must be treated with coffee and chocolate, preserves and sweetmeats, wines and dainties, jewels and pearle, gold and silver, silks and cosmetics, balls and assemblies, music and theatrical exhibitions: here you are satisfied with a little withered fungus!

Flora Lapponica, 1737

Tobaccos

The dry, cottony fibres that cover the leaves of mugwort and coltsfoot and make them inflammable enough to use as tinder (p. 151) have also led to their inclusion in smoking mixtures. Except for true tobaccos (*Nicotiana* spp. – native only in certain parts of north America, and not really within the scope of this book), coltsfoot, *Tussilago farfara*, has been the most popular temperate zone smoking leaf for more than two thousand years. Some of its oldest country names (Baccy plant, Poor man's baccy) testify to its traditional use; and even today it is the basic ingredient of most proprietary herbal smoking mixtures.

Given the choking and irritation normally associated with any kind of smoke, it is ironic that fumes from smouldering coltsfoot were first inhaled to ease coughs and asthma, and only later for pleasure. Pliny recommended burning the leaves on a fire of cypress wood, and inhaling the fumes through a reed (sipping a little wine between each puff – a foretaste of after-dinner indulgences to come!) It was probably during this period that the botanical name *Tussilago*, meaning 'cough-dispeller', was given to the plant.

Coltsfoot is an abundant weed of Europe and north America, growing equally happily on dry sands and water-logged loams, particularly where the ground has been disturbed. Its fine-floreted flowers, a little like yellow daisies, are amongst the first to appear in spring. They anticipate the leaves by some weeks, though these may begin to come through whilst the flower stalks are still standing. By June or July, which is the best time to pick the leaves, little evidence will be left of flowering save a few late seed clocks. It is possible to make an acceptable tobacco by simply drying the leaves for a couple of months and then flaking them finely with your fingers. But for a more distinguished product, try this recipe.

Hang the coltsfoot leaves in bunches in a warm, dry place until they are yellowish brown in colour, and have something of the consistency of dry chamois leather. Tear out the mid-ribs and pack the leaves tighly into a stone jar, adding a spoonful or two of brandy. Leave them in the jar for a month, then remove them in thin layers, roll them up and press the rolls under a weight for a few days. The result can be sliced and then crumbled like ordinary tobacco.

Coltsfoot gives a very easy, mild, pleasantly fragrant and probably harmless smoke, but it can be improved by the addition of other leaves: bearberry, bogbean, mullein, lavender and rosemary, eyebright and betony leaves, and the flowers of roses and chamomile have all been included in smoking mixtures. But beware the dried stems of old man's beard, *Clematis vitalba*, which are known as Smoking cane and Boy's bacca in south-west England. Unless these are completely dead and dry, they may contain irritant juices which cause ulceration on the lips and mouth. So powerful are these juices in the green plant, that professional beggars used to rub the leaves into scratches on their bodies, which raised large sores and enhanced the pitifulness of their appearance.

If, lastly, you should want herbal pipe-cleaner, remember the base of the flowering stems of purple moor-grass, *Molinia caerulea*, have been used in this way.

Tussocks

If you try to walk across a patch of open water in a fen or marsh, it is likely that you'll be forced to use clumps of tussock sedge, *Carex paniculata*, as stepping-stones. This distinctively tall species, which is reminiscent of pampas grass, is one of the first plants to invade shallow fresh water. The triangular-sectioned stems in each tuft are so closely packed that the clump is able to rise like an island above the water, and provide a base for other colonising plants like alder and willow. This also makes the tussocks pleasantly springy to walk on, and it is little wonder that in the Fens of

A tuft of tussock sedge (left), and a seat cut from a single piece of trunk in a Hertfordshire woodland. The forester allowed the natural flutings in the wood to form an ornamented headrest.

East Anglia, they were often cut off at the base and taken home for use as chimney-corner seats and church hassocks. Indeed the word hassock originally meant a tuft of grass.

On drier land, the natural resting-place for a tired walker is a disused ant-hill. These mounds raised by the meadow ant, *Lasius flavus*, are composed of finely crumbled soil, well drained and aerated, and are not only comfortably spongy to sit on (once the ants have moved on) but also tend to develop a cushion of the kind of mat-forming and fragrant herbs that can tolerate these light soils. So they are pleasing enough as seats just as they come, stuck out on a chalk down or sandy heath. But might they also have been the inspiration for those charming herb seats, carpeted with thyme and chamomile, that are such a feature of Elizabethan gardens?

Wicker

The plaiting of baskets from twigs and stalks is one of man's oldest crafts. It is widely believed to ante-date pottery, and probably suggested the idea of weaving cloth. In its simplest form it can be done without the use of any tools at all; yet no machine has been devised which can duplicate the process. Like all the most functional and attractive plantcraft, wicker baskets are built in response to the qualities of the raw materials, not imposed on them. Each wand is chosen individually. Each kink, each patch of bark colour, is incorporated where it will fit best. It is the uniqueness of each living twig that defeats mass production, and which gives plant-weaves their strength and liveliness.

In almost every civilisation the favourite wood for wickerwork has been willow. Its shoots have been made into beehives, cradles, chair-backs, lobster-pots, eel-traps and fruit bowls, but above all into baskets: baskets for bicycles, pets, laundry, wine-bottles, potatoes and picnics. There are few woods which can rival willow for pliancy and durability, or for lightness of weight – an important consideration when you are making containers for carrying about. And there are few kinds of tree which grow so readily and so fast, and which will tolerate repeated harvesting of their young shoots, sometimes every year.

In Europe the favourite species for wickerwork is the osier, *Salix viminalis*, a shrub which grows to about 4 metres in height, with long, straight branches and very narrow leaves. Some lowland villages still keep an osier bed in a damp corner, and it is managed in a similar way to a hazel coppice (p. 77). The bed is started by setting lengths of branch in the ground at regular intervals. Osiers spring up so willingly from cuttings (as do most willows) that it is possible to harvest some rough and ready wands within one year of planting. But these are not normally straight and regular enough for commercial use, and the bushes are cut back to the stool for three winters before a crop is taken off. After this period they can be cut annually for fifty or sixty years (John Evelyn reckoned until the 'World's end') and give a consistent crop of thirty or forty 2 metre wands per stool – which gives you some kind of measure of the shrub's prolific growth.

This kind of careful and regular husbanding produces the finest willow wands. But for small-scale work at home, almost any willow will do – including the weeping garden sorts. Look out for willows which have just been pollarded (the trimmimgs are often just dumped these days), or ask a landowner if you can crop one of his trees. Use secateurs, and choose the longest and straightest one-year old twigs. They can be cut any time between the autumn and spring.

Willow wands are traditionally treated in one of three

ways. 'Brown' willow is cut in the autumn and used with the bark intact. It is rougher in all senses than any other type of basket willow, and needs to be soaked in water for several days before use. 'Buff' willow is boiled for about five hours soon after cutting. The tannin and dye substances in the bark colour the wood a rich fawn, and the boiling kills off any potentially damaging fungi or insects. It also softens up the bark, which is stripped off some time later. 'White' osiers, which are peeled but uncooked wands, can only be obtained naturally during a short, critical period in the early spring. Most willow cut at this time will not peel unless it is boiled, as new wood is forming under the bark and binding it to the old. But there is a short period after the sap has begun to rise before new wood starts to form, when the bark, apparently lubricated by the spring juices, will slide off easily.

The rods are peeled by pulling them quickly through a sharp V-shaped metal notch held in a clamp (in some peasant cultures they use their teeth!). The rods are then stored, and just prior to weaving are thoroughly damped to make them pliant. The production of wickerwork is exceptionally simple in principle, but very complex in detail, and I would advise readers to go to specialist books if they are intending to try it themselves. The best work results from an almost instinctive understanding of the qualities and affinities of the raw material. Let me quote Dorothy Hartley where she is describing the dexterity of a weaver of eel traps:

'He has that look that comes to all men who work their wits against wild things, of thinking with their fingers. It is a curiously shaped funnel he is making, but its shape has evolved naturally between the flow of running water and the ways of eels – and of fen men. "Very queer things are eels," says the weaver, and the long green withys that twiddle and wave through the door of the shed or whisp along the floor seem to whisper back, "Yes, very queer things are eels – very strange ways".'

Made in England, 1939

The Paiute Indians made water jugs from fine willow twigs and, with an odd gesture of respect towards the people who drove them out of their homelands, based many of

Willow and water are natural familiars: Paiute Indian jugs, and a Fenland eel hive.

their designs on US Army canteens. But no white man would have thought of the refinement they built into these wicker jugs. They left the finest of gaps between each layer of withy, so that the jugs were not *quite* waterproof. Water would gradually seep out, dampen the outer surface, and, by evaporating in the sun, keep the remainder inside cool.

For sheer pleasure, they would also sometimes weave scented grasses amongst the withys, or twigs with different coloured barks. There are many kinds of willow whose barks are brightly coloured in winter. *Salix alba* var. *vitellina* is a beautiful pale orange, and there is a red-barked hybrid that in Tewkesbury is used to bind up bundles of spring onions and celery, to emphasise their whiteness.

Many other whippy twigs have been employed in basket-weaving, and can be used in much the same way as withys, or mixed in with them to give multi-coloured wickerwork: bramble, wild clematis, dogwood (bright crimson in winter), elm, honeysuckle, snowberry, ivy, hazel, lime, privet, rose, broom (bright green in the spring).

There is no need to debark these twigs unless you are making articles to last, and by mixing different species together you will be able to produce baskets patterned by the natural colours of the woods. And for really eccentric, throwaway items why not leave on buds, flowers and all.

Epilogue

The ancient working relationship between men and plants that I have sketched in this book is not a closed story. New uses, new transformations, will continue to be discovered for as long as we look at plants with respect and imagination. Even as I write this, the first commercial plantation of jojoba, *Simmondsia chinensis*, is being set down in the United States. Jojoba is a small shrub that grows wild and native in the desert regions of Mexico and south-western North America, but it is only recently that its seeds have been found to contain an oil which shares with sperm oil a peculiar indifference to high temperatures and pressures. The automobile industry says that such an oil is essential for the lubrication of transmissions systems, and up till now has stubbornly insisted that the oil of sperm whales is the only remotely suitable material. So the genocide of those vast, gentle creatures has continued. Now – though it is a slender hope – the economic pressures that underpin their slaughter may be eased a little by the good offices of a desert weed.

A seed oil which eases the progress of the modern automobile; it is one kind of vegetable victory I suppose, though not without its ironies. But the potentialities of plants are so limitless that perhaps it is not too much of a fantasy to look forward to an era of wooden carriages powered by 'the essence of stone' uncovered by those Australian chemists.

Bibliography

Sources, References and Further Reading

The majority of the historical works in this list contain first-hand accounts of the traditional uses of wild plants. The modern works often contain detailed descriptions of the craft practices themselves – dyeing, rushwork, etc – and this is usually clear from their titles. These would be useful first books for those wishing to pursue a particular craft in more detail.

ALLABY, Michael, *The Survival Handbook*, Macmillan, 1975
ARNOLD, James, *The Shell Book of Country Crafts*, John Baker, 1968
ASSINIWI, Bernard, *Survival in the Bush*, Copp Clark, 1972
BADHAM, C. D., *On the Esculent Funguses of England*, 1863
BOLTON, Eileen, *Lichens for Vegetable Dyeing*, Studio Vista, 1972
BUCHMAN, Dian Dincin, *Feed Your Face*, Duckworth, 1973
CALDER, Ritchie, *The Life Savers*, Pan, 1961
CASTLETON THOMAS, Virginia, *Secrets of Natural Beauty*, Harrap, 1973
CHAPMAN, V. J., *Seaweeds and their Uses*, Methuen, 2nd Edition 1970
CLARK, Robin & HINDLEY, Geoffrey, *The Challenge of the Primitives*, Jonathan Cape, 1975
COBBETT, William, *Cottage Economy*, 1823
COBBETT, William, *Woodlands* 1825
Country Bazaar, Architectural Press, 1974, Fontana, 1976
CRAUFORD, Lady Ruth, *Country Women at War*, published by author, 1968
CRIPPS, Ann, ed., *Rescuing the Past*, David & Charles, 1973
CULPEPPER, Nicholas, *Complete Herbal*, 1853 edition
DAKER, Jerry, *Second Back to Nature Almanac*, Simon & Schuster, 1974

DENSMORE, Frances, *How Indians Use Wild Plants*, Bureau of American Ethnology, 1928, Dover, 1974

DOUGLAS, J. Sholto, & HART, Robert A. de J., *Forest Farming*, Watkins Books, 1976

Dye Plants and Dyeing, A Handbook, Brooklyn Botanic Garden, 1964

EASTWOOD, Dorothea, *Mirror of Flowers*, Derek Verschoyle, 1953

EATON, Allen H., *Handicrafts of the Southern Highlands*, Russell Sage Foundation, 1937, Dover, 1973

EDLIN, H. L., *British Plants & Their Uses*, Batsford, 1951

EDLIN, H. L., *Man and Plants*, Aldus Books, 1967

EDLIN, H. L., *Woodland Crafts of Britain*, David and Charles, 1973

ELLIS, E. A., *The Broads*, Collins, 1965

EVELYN, John, *Sylva, A discourse of Forest-Trees*, 1664

FERRY, B. W., BADDELEY, M. S., HAWKESWORTH, D. L. (ed), *Air Pollution and Lichens*, Athlone Press, 1973

FLORANCE, Norah, *Practical Rushwork*, Dryad, 1972

FORSYTH, A. A., *British Poisonous Plants*, for the Ministry of Agriculture, Fisheries & Food, HMSO, 1968

GENDERS, Roy, *A History of Scent*, Hamish Hamilton, 1972

GENDERS, Roy, *The Scented Wild Flowers of Britain*, Collins, 1971

GERARD, John, *The Herball or Generall Historie of Plantes*, enlarged and amended by Thomas Johnson, 1633

GIBBONS, Euell, *Stalking the Blue-eyed Scallop*, David McKay, 1964

GIBBONS, Euell, *Stalking the Healthful Herbs*, David McKay, 1966

GIBBONS, Euell, *Stalking the Wild Asparagus*, David McKay, 1962

GILMOUR, J. & WALTERS, S. M., *Wild Flowers*, Collins, 1954

GIMINGHAM, C. H., *Ecology of Heathlands*, Chapman and Hall, 1972

GRIEVE, Mrs. M., *A Modern Herbal*, 1931 Penguin, 1976

GRIGSON, Geoffrey, *A Herbal of All Sorts*, Phoenix House, 1959

GRIGSON, Geoffrey, *Dictionary of English Plant Names*, Allen Lane, 1974

GRIGSON, Geoffrey, *Gardenage*, Routledge and Kegan Paul, 1952

GRIGSON, Geoffrey, *The Englishman's Flora*, Phoenix House, 1958

GRIGSON, Geoffrey, *The Shell Country Book*, Dent, 1962

HADFIELD, Miles, *British Trees*, Dent, 1957

HALE, Thomas, *Compleat Body of Husbandry*, London, 1756

HARDY, Thomas, *The Woodlanders*, 1887

HARRIS, Ben Charles, *Eat the Weeds*, Barre, 1972

HARTLEY, Dorothy, *Food in England*, Macdonald, 1954

HARTLEY, Dorothy, *Made In England*, Eyre Methuen, 1939

HATFIELD, Audrey Wynne, *How to Enoy Your Weeds*, Muller, 1969

HAYES, M. Vincent, *Artistry in Wood*, David and Charles, 1973

HEMPHILL, Rosemary, *Herbs and Spices*, Penguin, 1966

HEMPHILL, Rosemary, *Herbs for All Seasons*, Penguin, 1972

HERITEAU, Jacqueline, *Potpourris and Other Fragrant Delights*, Simon and Schuster, 1972 Lutterworth, 1975,

HILLS, Lawrence D., *Pest Control Without Poisons*, Henry Doubleday Research Association, 1964

HOPPE, H., *Whittling and Wood Carving*, Sterling Publishing Co, New York, 1969

HUTCHINGS, Margaret, *Nature's Toyshop*, Mills and Boon, 1975

HUXLEY, Anthony, *Plant and Planet*, Allen Lane, 1974

HVASS, Else, *Plants that Serve Us*, Blandford, 1960

HYAMS, Edward, *Plants in the Service of Man*, Dent, 1971

JACKSON, Nora & PENN, Philip, *A Dictionary of Natural Resources*, Pergamon, 1969

JAMES, George Wharton, *Indian Basketry*, 1909, Dover Edition, 1972

JENKINS, J. Geraint, *Traditional Country Craftsmen*, Routledge and Kegan Paul, 1965

JOHNSON, Charles, *The Useful Plants of Great Britain*, 1862

JOHNSON, Hugh, *The International Book of Trees*, Mitchell Beazley, 1973

JONES, John L., *Crafts from the Countryside*, David and Charles, 1975

KEEN, Barbara & ARMSTRONG, Jean, *Herb Gathering*, Brome and Schimmer, 1942

KENNETT, Frances, *A History of Perfume*, Harrap, 1975

KNOCK, A. G., *Willow Basket-work*, Dryad, 1970

KREIG, Margaret B., *Green Medicine*, Harrap, 1965

LANDSBOROUGH, Rev. D., *Popular History of British Seaweeds*, 1857

LAUREL, Alicia Bay, *Living on the Earth*, Wildwood House, 1971

Bibliography 169

LESCH, Alma, *Vegetable Dyeing*, Watson-Guptill Publications, 1970

LEVY, Juliette de Bairacli, *Herbal Handbook for Everyone*, Faber and Faber, 1966

LI, C. P., *Chinese Herbal Medicine*, US Department of Health, Education and Welfare, 1974

LOEWENFELD, Claire & BACK, Philippa, *The Complete Book of Herbs and Spices*, David and Charles, 1974

LOEWENFELD, Claire, *Nuts*, Faber, 1957

LOVELOCK, Yann, *The Vegetable Book*, Allen and Unwin, 1972

MABEY, Richard, *The Pollution Handbook*, Penguin, 1974

MANNERS, J. E., *Country Crafts Today*, David and Charles, 1974

MEYER, Joseph E., *The Herbalist*, 1918

MILLER, Philip, *Gardeners Dictionary*, 1741 ed.

MONCRIEFF, Robert, *Chemical Senses*, Leonard Hill, 1967

MORRIS, M. G. & PERRING, F. H., (ed) *The British Oak*, Botanical Society of the British Isles and E. W. Classey, 1974

PAPANEK, Victor, *Design for the Real World*, Thames and Hudson, 1972

PARKINSON, John, *Theatrum Botanicum*, 1640

PATURI, Felix R., *Nature, Mother of Invention*: The Engineering of Plant Life, Thames & Hudson 1976

PERRING, F. H., SELL, P. D., WALTERS, S. M., *Flora of Cambridgeshire*, Cambridge University Press, 1964

PETCH, C. P., SWANN, E. L., *Flora of Norfolk*, Jarrold, 1968

PITT, Robert, *The Crafts and Frauds of Physick Exposed*, 1702

POLLARD, HOOPER, MOORE, *Hedges*, Collins 1974

PORTA, John Baptista, *Natural Magick*, 1658

QUELCH, Mary Thorn, *Herbs for Daily Use*, Faber, 1941

RACKHAM, Oliver, *Hayley Wood, Its History and Ecology*, Cambridgeshire and Isle of Ely Naturalists' Trust, 1975

RAMSBOTTOM, John, *Mushrooms and Toadstools*, Collins, 1953

RANSON, F., *British Herbs*, Penguin Books, 1949

RICHARDSON, David, *The Vanishing Lichens*, David and Charles, 1975

ROBERTSON, Seonaid M., *Dyes from Plants*, Van Nostrand, 1973

ROHDE, Eleanour Sinclair, *A Garden of Herbs*, The Medici Society, 1936

ROHDE, Eleanour Sinclair, *The Old English Herbals*, Minerva Press edition, 1972

ROHDE, Eleanour Sinclair, *The Scented Garden*, The Medici Society, 1931

SALISBURY, Sir Edward, *Weeds and Aliens*, Collins, 1961

SALISBURY, William, *The Botanist's Companion and the Uses of Plants*, 1816

SINCLAIR, George, *Hortus Gramineus Woburniensis*, 1824

SMITH, Alexander H., *Mushroom Hunter's Field Guide*, University of Michigan Press, 1958

SOPER, Tony, *The Shell Book of Beachcombing*, David and Charles, 1972

STEVENSON, Violet (ed), *A Modern Herbal*, Octopus, 1974

TAYLOR, Norman, *Plant Drugs that Changed the World*, Allen and Unwin, 1965

THURSTAN, Violetta, *The Use of Vegetable Dyes*, Dryad, 1972

TREASE, G. E., *A Textbook of Pharmacognosy*, Bailliere, Tindall and Cox, 1961

TURNER, William, *The Herbal*, 1568 edition

TUSSER, Thomas, *Five Hundred Pointes of Good Husbandrie*, 1573

VOGEL, Virgil J., *American Indian Medicine*, University of Oklahoma Press, 1970

WALLIS, T. E., *Textbook of Pharmacognosy*, Churchill, 1960

WEINER, Michael A., *Earth Medicine – Earth Foods*, Collier Macmillan, 1972

WHEAT, Margaret, *Survival Arts of the Primitive Paiutes*, University of Nevada Press, 1967

WHITE, Gilbert, *The Natural History of Selborne*, 1789

WILKINSON, Gerald, *Trees in the Wild*, Stephen Hope, 1973

WITHERING, William, *An Arrangement of British Plants*, 7th ed. 1830

WYATT, John, *The Shining Levels*, Geoffrey Bles, 1973

Index of plants